BOOLEAN LOGIC, EXPRESSIONS AND THEORIES

AN OVERVIEW

THEORETICAL AND APPLIED MATHEMATICS

Additional books and e-books in this series can be found on Nova's website under the Series tab.

THEORETICAL AND APPLIED MATHEMATICS

BOOLEAN LOGIC, EXPRESSIONS AND THEORIES

AN OVERVIEW

VICTORIA C. CARLSEN
EDITOR

nova
science publishers
New York

NOTICE TO THE READER

Library of Congress Cataloging-in-Publication Data

ISBN: 978-1-53616-985-0

Published by Nova Science Publishers, Inc. † New York

CONTENTS

PREFACE

The Boolean function network is a systematical approach proposed for the inference of gene regulatory networks and related Boolean functions. This procedure utilizes two steps to integrate the hidden Markov model, likelihood ratio test and Boolean functions for discovering direct pairwise relations between genes from time-course transcriptome data. In this compilation, the authors justify the need for this novel approach and describe the inference procedure.

Next, an extended Boolean logic is introduced, denoted by LLT, called starfinite or hyperfinite logic. LLT is closely related to infinitary logics, which have been investigated extensively.

Lastly, generalized Boolean functions are introduced, and an overview with regard to constructions of Golay complementary sequences from generalized Boolean functions is given.

Boolean function network (BFN) is a systematical approach proposed to the inference of gene regulatory networks (GRNs) and the related Boolean functions. This procedure utilizes two steps to integrate hidden Markov model (HMM), likelihood ratio test and Boolean functions for discovering direct pairwise relations between genes from time-course transcriptome data. The low computational complexity of BFN makes it advantageous for the applications on the genome-wide scale. In Chapter 1 the authors justify the need for novel approach and describe the inference procedure.

As explained in Chapter 2, classical propositional logic can be regarded as a logic equipped with the finite disjunctions $\bigvee_{i=1}^{n}$ and conjunctions $\bigwedge_{i=1}^{n}$ for every natural number $n \geq 2$ (in addition to the negation $\neg$) – instead of the usual binary disjunction $\bigvee$ and conjunction $\bigwedge$ only. This variant of the classical propositional logic can be extended so that instead of the naturals n, *hyperfinite* naturals (i.e., *hypernaturals*) N are choosen, i.e., the operations $\bigvee_{i=1}^{N}$ and $\bigwedge_{i=1}^{N}$ are assumed. This extended *Boolean logic* is introduced in Chapter 2, denoted by $L_T^{\mathcal{L}}$, called "star-finite", or "hyperfinite" logic ("star-finite" means in non-standard analysis an "infinite" whose behaviour is "like the finite"). The logic $L_T^{\mathcal{L}}$ is closely related to *infinitary logics*. Infinitary logics have been investigated extensively. It is proven that the star-finite logic introduced here is even stronger, in a sense, than infinitary logics, in general.

Classical proposition logic has *Boolean algebra* as its associated concept of algebraization. It is shown that with star-finite logic the non-standard enlargement of Boolean algebra can be associated. It is proven that such an extended Boolean algebra is isomorphic to a Boolean set algebra closed under the hyperfinite unions and intersections. This is remarkable because Boolean γ-algebras are not representable by Boolean γ-set algebras, if γ is an infinite cardinality. Also a new version of the classical Stone theorem is proved: Boolean algebras are isomorphic to some Boolean set algebra with a *hyperfinite* unit.

As shown in Chapter 3, a generalized Boolean function (GBF) is a mapping from direct product Z_2^m to set Z_H, where $Z_2 = \{0, 1\}, Z_H = \{0, 1, 2, \cdots, H-1\}$, and integers $m(\geq 1), H(\geq 2)$. GBFs can be divided into standard and non-standard ones. GBFs play fairly important roles in constructing H phase shift keying (PSK) Golay complementary sequences (GCSs). In particular, binary phase-shift keying (BPSK) or quadrature phase-shift keying (QPSK) GCSs can be resulted in for $H = 2$ or 4, respectively. For given m, standard H-PSK GCSs of length $N = 2^m$ have $m!H^{m+1}/2$ $(m, H \geq 2)$, and $(n-2)!(n-2)4^n$ for non-standard QPSK GCSs $(n = m + 3, m \geq 1)$. GCSs are closely associated with the control of peak envelope power (PEP) of transmitted signals in an orthogonal frequency-division multiplexing (OFDM) communication system. It has been proved that upper bound of peak-to-mean envelope power ratio (PMEPR) of transmitted signals in an OFDM communication system, employs BPSK or QPSK GCSs, does not excelled 2. Further, based on BPSK or QPSK GCSs referred to above, quadrature amplitude modulation (QAM) GCSs can be designed. In general, for an OFDM communication system employing resultant 4^q-QAM GCSs of length $N = 2^m$, upper bound of PEP of transmitted signals of this system can be controlled not to excel $\frac{6N(2^q-1)}{2^q+1}$.

In: Boolean Logic, Expressions and Theories   ISBN: 978-1-53616-985-0
Editor: Victoria C. Carlsen   © 2020 Nova Science Publishers, Inc.

Chapter 1

BOOLEAN FUNCTION NETWORKS

Maria Simak[1,2], Henry Horng-Shing Lu[2,],*
Chen-Hsiang Yeang[3] and Jinn-Moon Yang[1]

[1]Institute of Bioinformatics and Systems Biology,
National Chiao Tung University, Hsinchu, Taiwan, ROC
[2]Institute of Statistics, National Chiao Tung University,
Hsinchu, Taiwan, ROC
[3]Institute of Statistics, Academia Sinica, Taipei, Taiwan, ROC

ABSTRACT

Boolean function network (BFN) is a systematical approach proposed to the inference of gene regulatory networks (GRNs) and the related Boolean functions. This procedure utilizes two steps to integrate hidden Markov model (HMM), likelihood ratio test and Boolean functions for discovering direct pairwise relations between genes from time-course transcriptome data. The low computational complexity of BFN makes it advantageous for the applications on the genome-wide scale. In this chapter we justify the need for novel approach and describe the inference procedure.

* Corresponding Author's E-mail: hslu@stat.nctu.edu.tw.

INTRODUCTION

The gene expression is a complex multilevel process which when put in simple terms goes from transcription of DNA to mRNA and further to translation of mRNA into proteins. The collection of DNA, mRNA, proteins and other macromolecules that are involved in this process, together with their interactions within cell, constitute so called gene regulatory network (GRN). Notably, aside from external signals which govern the course of gene expression, some proteins also play the role of regulators by binding to promoter region of specific genes and either enabling the RNA polymerase to engage with DNA strand and initiate transcription process or on contrary, preventing it. These proteins are called transcription factors (TF), and their relations with protein-coding genes are crucial for understanding of control of gene expression. Measuring the abundance of mRNA is one of the ways to observe and make conclusion about regulatory processes within cell. At the genome-wide scale it became possible only quite recently with help of high throughput experimental technologies: microarrays and RNA-seq. The former is more affordable while the last one is typically more precise and able to identify potentially novel genes. There are two major experimental designs to identify potential regulatory targets: perturbation (knockout or overexpression) and observational time-series. While the first one can identify regulatory targets of specific gene quite precisely, the number of experiments that are required makes it not applicable on genome scale. Thus the main focus of this chapter will be at the exploration of the time-course expression data.

The task of GRN inference or reverse-engineering is one of the major challenges in modern bioinformatics and systems biology and has given rise to many approaches to tackle this challenge. In Table 1 we have summarized standard methods in this field and their characteristics such as: ability to distinguish between direct and indirect interactions, ability to grasp a dynamic of the data, ability to identify directedness of relation and ability to assign function to relation between genes. Methods are arranged in order of increasing complexity and to some degree of increasing accuracy. Such methods as Boolean network and dynamic Bayesian network most of the time require applying certain heuristics to make them applicable on the genome-wide scale, which leads to loosing accuracy. Moreover, the performance of method depends greatly on the dataset i.e., number of available observations and presence of noise. Therefore, sometimes a simple but robust method can outperform more complex one on a given real dataset. Table 1 does not include another big class of GRN inference methods, which are based on differential equations, since they are best in describing molecular kinetics of cell in detail, yet are not pertinent to the genome size datasets.

As a summary of Table 1, in comparison to other existing approaches, the Boolean function network(BFN) method is relatively simple (we discuss its complexity below), has competitive accuracy (as it was demonstrated in [6]), provides direct links, captures dynamic of time-series data, but what is more important it assigns optimal Boolean function, time-delay and direction to every gene relation in the network.

Table 1. GRN reverse engineering approaches and their characteristics

Type of approach	Example	Establishes direct links	Captures dynamic of the data	Establishes directionality of the links	Identifies functional relation	Complexity
Correlation networks	WGCNA [1]	No	No	No	No	$O(n)$
Information theory based methods	Relevance networks [2]	No	No	No	No	$O(n^2)$
	MRNET [3]	Yes	No	No	No	$O(n^2)$
	ARACNE [4]	Yes	No	No	No	$O(n^2)$
	CLR [5]	Yes	No	No	No	$O(n^2)$
Boolean function network (BFN)	BFN [6]	Yes	Yes	Yes	Yes	$O(n)$ for Test1 $O(n^3)$ for both Test1 and Test2
Regression-based methods	GENIE3 [7]	Yes	Yes	Yes	No	$O(n^2)$
Graph-based methods	GeneNet [8]	Yes	Yes	Yes	No	$O(n^3)$
Boolean networks (BN)	BoolNet [9]	Yes	Yes	Yes	Yes	$O(2^n)$
Dynamic Bayesian networks	G1DBN [10]	Yes	Yes	Yes	No	$O(2^n)$

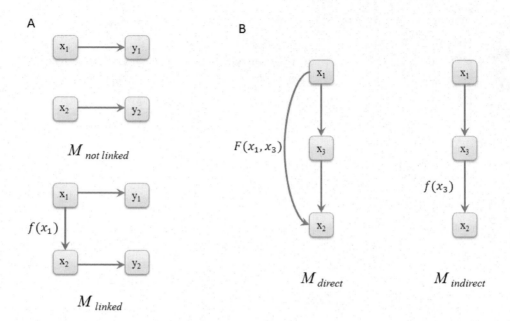

Figure 1. Graphical models of Test1 and Test2: A) Not-linked model vs linked model. In the model M$_{\text{not linked}}$ gene *One* and gene *Two* are unrelated; whereas in the model M$_{\text{linked}}$ gene *One* regulates gene *Two*. For both models, x_1 and x_2 are hidden gene states and y_1 and y_2 are their corresponding observed values. B) Direct model vs indirect model. In the direct model M$_{\text{direct}}$, both gene *One* and gene *Three* regulate gene *Two* directly; while in the indirect model M$_{\text{indirect}}$, gene *One* regulates gene *Two* through intermediate gene *Three*. In figure B we omit depicting observed values y_i for the sake of simplicity.

BOOLEAN FUNCTION NETWORK METHOD

The BFN method is a two-step procedure, which includes an identification of pairwise dependencies between genes by Test 1, followed by verification whether those links are direct by Test 2. Both Test1 and Test 2 are based on the comparison of likelihoods of two alternative models. Test 1 serves to find an optimal Boolean function and time delay between two genes, which would maximize the likelihood ratio of a model with a link over a model without a link. Figure 1(A) provides the graphical illustration of Test 1. Analogically, Test 2, illustrated by Figure 1(B), maximizes the likelihood ratio of a direct model over indirect model. Suppose, we have detected the following links $x_1 \rightarrow x_3$,

$x_3 \rightarrow x_2$, $x_1 \rightarrow x_2$ by Test 1. As it is shown at Figure 1(B) there are two ways of gene regulation in this case. In the first scenario, both x_1 and x_3 regulate x_2 directly. In the second scenario, x_1 has indirect effect on x_2 through x_3. Therefore the purpose of Test 2 is to establish whether link $x_1 \rightarrow x_2$ is direct and eliminate it in case if it is not.

Discretization

The expression profile is the measured abundance of mRNAs for each determined point in time. As we have mentioned earlier, the source for this type of data is usually either microarray or next generation sequencing experiment. We assume that variables (genes) are arranged horizontally and n is the number of genes. The observations (time-points) are arranged vertically with the total number of columns is m, n $\gg$ m. Naturally, the range of values varies greatly from one gene to another. In order to enable comparison of the expression profiles of different genes, the expression values have to be standardized to the same scale, i.e., converted to the standard range of (0, 1] for every gene. For this purpose we apply empirical cumulative distribution function (ECDF) transformation, which can be illustrated with the following toy example.

Let's assume we are given next expression values for a gene:

$$\{5; 2; 1; 1; 0.05; 10; 50\}.$$

To the every expression value, we assign the index number divided by the total number of observations. When a tie occurs, the index number is incremented by the number of tied observations; and it is the same for all tied observations. The result of transformation will be as follows:

$$\{\frac{5}{7}; \frac{4}{7}; \frac{3}{7}; \frac{3}{7}; \frac{1}{7}; \frac{6}{7}; 1\}.$$

Boolean Network and Boolean Functions

We define the Boolean network as a set of vertices $V = \{x_1 \ldots x_n\}$ representing genes and a set of all unary and binary Boolean functions $f = \{f_1 \ldots f_6, F_1 \ldots F_{42}\}$, which represent relations between nodes.

Boolean function is a mapping of the form $f: B_k \rightarrow B$, where $B = \{0, 1\}$ is a Boolean domain and k is arity of the function. For every k there exist a finite set of non-trivial Boolean functions which can be represented in the form of truth table. In Table 2 and 3, we enumerate all possible non-trivial Boolean unary and binary functions correspondingly.

Besides all possible non-trivial Boolean functions with unique definite output, we also consider functions with two possible outputs, $\{0, 1\}$, which means that either 0 or 1 may appear in the output for the same input assignment. In Table 2, function f_1 (equivalence) represents upregulation of gene x_2 by gene x_1. Function f_2 (negation) stands for downregulation of gene x_2 by gene x_1. Functions f_3 and f_4 reflect the relation of necessity and its negation correspondingly. That is, function f_3 explains condition "gene x_2 cannot be turned on unless gene x_1 is turned on" and function f_4 states opposite "gene x_2 cannot be turned off unless gene x_1 is turned on". Functions f_6 express sufficiency and f_5 is its negation. If gene x_1 is sufficient for gene x_2 it means that knowing that gene x_1 is on we can claim that gene x_2 is on as well. However, it is not legit to assert that if gene x_1 is off then gene x_2 is off too. Whilst function f_5 represent the statement "if gene x_1 is on then gene x_2 must be off".

Figure 2 illustrates why we went beyond the simple Boolean functions of equivalence and negation in attempt to describe gene relations. It depicts four relations that were revealed as a result of applying BFN method to the whole

Table 2. Truth table for unary functions f_1-f_6. x_1 is input of function (source gene) and x_2 is output of function (target gene)

x_1	x_2					
	f_1	f_2	f_3	f_4	f_5	f_6
0	0	1	0	1	{0,1}	{0,1}
1	1	0	{0,1}	{0,1}	0	1

Table 3. Truth table for binary functions F_1–F_{42}. x_1 and x_2 are input of the function and x_3 is output

x_1	x_2		F1	F2	F3	F4	F5	F6	F7	F8	F9	F10	F11	F12	F13	F14
												x_3				
0	0		{0,1}	{0,1}	0	0	0	1	1	1	1	1	{0,1}	{0,1}	{0,1}	{0,1}
0	1		0	0	1	1	1	0	0	0	1	1	0	0	0	0
1	0		0	1	0	1	1	0	0	1	0	1	0	0	1	1
1	1		1	0	0	0	1	0	1	1	1	0	0	1	0	1

x_1	x_2		F15	F16	F17	F18	F19	F20	F21	F22	F23	F24	F25	F26	F27	F28
												x_3				
0	0		{0,1}	{0,1}	{0,1}	{0,1}	0	0	0	0	1	1	1	1	0	0
0	1		1	1	1	1	{0,1}	{0,1}	{0,1}	{0,1}	{0,1}	{0,1}	{0,1}	{0,1}	0	0
1	0		0	0	1	1	0	0	1	1	0	0	1	1	{0,1}	{0,1}
1	1		0	1	0	1	0	1	0	1	0	1	0	1	0	1

x_1	x_2		F29	F30	F31	F32	F33	F34	F35	F36	F37	F38	F39	F40	F41	F42
												x_3				
0	0		0	0	1	1	1	1	0	0	0	0	1	1	1	1
0	1		1	1	0	0	1	1	0	0	1	1	0	0	1	1
1	0		{0,1}	{0,1}	{0,1}	{0,1}	{0,1}	{0,1}	0	1	0	1	0	1	0	1
1	1		0	1	0	1	0	1	{0,1}	{0,1}	{0,1}	{0,1}	{0,1}	{0,1}	{0,1}	{0,1}

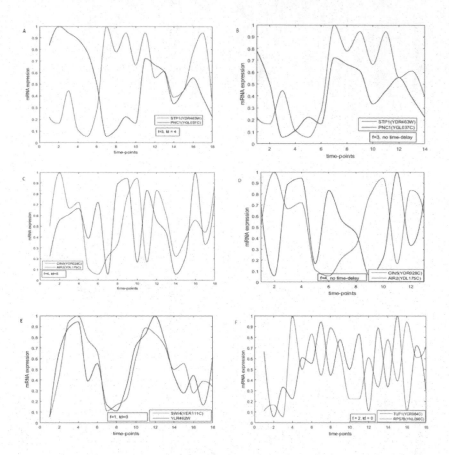

Figure 2. Examples of f_1, f_2, f_3 and f_4 functions that were inferred from the whole genome *S. cerevisiae* dataset. (A) *STP1* regulates *PNC1* and their relation is described by the function f_3 with time-delay equals to 4. (B) Expression profile of *PNC1* is shifted by 4 time-points . (C) *CIN5* regulates *AIR2* and their relation is described by the function f_4 with time-delay equals to 5. (D) Expression profile of *AIR2* is shifted by 5 time-points. (E) *SWI4* and *YLR462W* are co-expressed: f_1, time-delay equals 0. (F) *TUP1* and *RPS7B* have opposite expression profiles: f_2, time-delay equals to 0. For the purpose of visualization, in this figure we used already discretized transcriptome values.

genome yeast dataset of Spellman et al. [6, 11] and were validated with Saccharomyces genome database (SGD) [12]. Figure 2(A) is the original picture of expression of genes *STP1* and *PNC1*, which were found to be related with function f_3 and time-delay equals to 4. Figure 2(B) is a corresponding image where expression profile of target gene *PNC1* was shifted so that time-delay between genes was removed for the illustrative purpose. Similarly, Figure 2(C) demonstrates genes *CIN5* and *AIR2*, whose relation was found to be described with f_4 function and time-delay equals to 5. Figure 2(D) is a corresponding image of genes *CIN5* and *AIR2* with shifted profile of the target gene *AIR2*. Indeed, from Figure 2(B) it can be seen that while genes *STP1* and *PNC1* have similar profile with overall correlative trend, there are significant divergences at points 3, 5 and 11. Analogously, profiles of genes *CIN5* and *AIR2* are opposite most of the time with few divergences. For comparison, Figures 2(E) and 2(F) are examples of functions f_1 (equivalence) and f_2 (negation), that were inferred along with relations pictured at Figures 2A and 2C. Obviously, the profiles of genes *SWI4* and *YLR462W* show much better correlation, same as genes *TUP1* and *RPS7B* are more clearly oppose to each other. Therefore this example demonstrates that expanding set of Boolean functions beyond simple equivalence or negation endorse capturing more complex expression patterns and it is also beneficial when dealing with noisy dataset.

Functions F_1-F_{42} with binary inputs may or may not have simple and intuitive forms. For instance, F1 realizes an "AND" gate of two inputs; yet for instance F_{24} does not have a simple Boolean functional form. In this study, we are primarily concerned with finding pairwise dependencies in the gene network. The Boolean functions with binary inputs are auxiliary in determining the directness of links.

Test 1

In order to identify the pairwise dependencies between genes, we examine two models for every possible pair of genes. One model represents the situation where genes are linked and the other model suggests there is no link between the genes under consideration.

Assume, that $y_1(t)$ and $y_2(t)$ are continuous observed values of gene *One* and gene *Two* at the time point t respectively. The notations of $x_1(t)$ and $x_2(t)$ are the corresponding discrete latent variables. The notation of τ represents the time-delay between genes. Two competitive models are shown at Figure 1(A).

In order to establish which one of the models explains data better, we use the likelihood ratio:

$$R = \frac{L_{linked}}{L_{not\ linked}}.$$

The larger this ratio is, the more significant the link is. The likelihoods of models can be written respectively:

$$L_{not\ linked} = \prod_t \sum_{\substack{x_1(t)\\x_2(t)}} P(x_1(t)) \cdot P(x_2(t+\tau)) \cdot P(y_1(t)|x_1(t)) \cdot P(y_2(t+\tau)|x_2(t+\tau)), \quad (1)$$

$$L_{linked} = \prod_t \sum_{\substack{x_1(t)\\x_2(t)}} P(x_1(t)) \cdot P(x_2(t+\tau)|x_1(t)) \cdot P(y_1(t)|x_1(t)) \cdot P(y_2(t+\tau)|x_2(t+\tau)), \quad (2)$$

where the product is taken over all time points; at each time point t the likelihood score is marginalized over all possible latent variable states of $x_1(t)$ and $x_2(t)$.

According to Bayes' theorem, $P(y_k(t)|x_k(t)) = \frac{P(x_k(t)|y_k(t))P(y_k(t))}{P(x_k(t))}.$ (3)

When $P(y_1(t)|x_1(t))$ and $P(y_2(t)|x_2(t))$ in formulas (1) and (2) are replaced with (3), we obtain the followings:

$$L_{not\ linked} = \prod_t \sum_{\substack{x_1(t)\\x_2(t)}} \frac{\begin{array}{c} P(x_1(t)) \cdot P(x_2(t+\tau)) \cdot \\ P(x_1(t)|y_1(t)) \cdot P(y_1(t)) \end{array}}{P(x_1(t))} \cdot \frac{P(x_2(t+\tau)|y_2(t+\tau)) \cdot P(y_2(t))}{P(x_2(t+\tau))}$$

$$=$$

$$= \prod_t \sum_{\substack{x_1(t)\\x_2(t)}} P(x_1(t)|y_1(t)) \cdot P(y_1(t)) \cdot P(x_2(t+\tau)|y_2(t+\tau)) \cdot P(y_2(t)) =$$

$$= \prod_t P(y_1(t)) \cdot P(y_2(t)) \sum_{\substack{x_1(t)\\x_2(t)}} P(x_1(t)|y_1(t)) \cdot P(x_2(t+\tau)|y_2(t+\tau)),$$

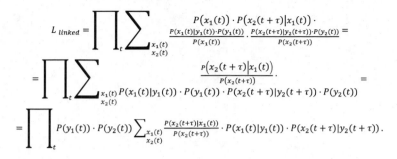

The terms of $P(y_1(t))$ and $P(y_2(t))$ are taken outside of the sum in both formulas because $P(y_k(t))$ is constant as $y_k(t)$ is the realization of random variable $x_k(t)$. They will be cancelled out in likelihood ratio and they can be omitted in formulas for $L_{not\ linked}$ and L_{linked}. So the formulas can be rewritten as next:

$$L_{not\ linked} \propto \prod_t \sum_{\substack{x_1(t) \\ x_2(t)}} P(x_1(t)|y_1(t)) \cdot P(x_2(t+\tau)|y_2(t+\tau)),\quad (4)$$

$$L_{linked} \propto$$

$$\prod_t \sum_{\substack{x_1(t) \\ x_2(t)}} \frac{P(x_2(t+\tau)|x_1(t))}{P(x_2(t+\tau))} \cdot P(x_1(t)|y_1(t)) \cdot P(x_2(t+\tau)|y_2(t+\tau)).\quad (5)$$

The estimate of conditional probability $P(x_k(t)|y_k(t))$ is the empirical CDF $\hat{F}_k(t)$, thus:

$$P(x_k(t) = 1|y_k(t)) = \hat{F}_k(t),$$

$$P(x_k(t) = 0|y_k(t)) = 1 - \hat{F}_k(t).$$

For simplicity, we will use $p_{x1x2}{}^t$ notation instead of product $P(x_1(t)|y_1(t)) \cdot P(x_2(t+\tau)|y_2(t+\tau))$.

For example, $p_{00}{}^t = P(x_1(t) = 0|y_1(t)) \cdot P(x_2(t+\tau) = 0|y_2(t+\tau))$.

Note that $P(x_k(t))$ is a marginal probability and it can be computed as follows:

$$q_{0k} = P(x_k(t) = 0) = \frac{\sum_t P(x_k(t)=0|y(t))}{m},$$

$$q_{1k} = P(x_k(t) = 1) = \frac{\sum_t P(x_k(t)=1|y(t))}{m}.$$

The conditional probability of Boolean state of variable $x_2(t + \tau)$ given $x_1(t)$ in (5) becomes:

$$P(x_2(t + \tau)|x_1(t)) =$$

$$\begin{cases} \delta(x_2(t + \tau) = f(x_1(t))), \text{ if } f(x_1(t)) \neq \Omega; \\ P(x_2(t + \tau)), \text{ if } f(x_1(t)) = \Omega; \end{cases} \Omega = \{0 \text{ or } 1\}. \qquad (6)$$

Equation (6) specifies the pattern for each of the six possible Boolean functions of one variable. If it is f_1 (equivalence), then p_{00} and p_{11} are given weight 1, while p_{01} and p_{10} are set to zero in accordance to the truth table of f_1. For computation reason in practice, we will use ε and $1 - \varepsilon$ instead of 0 and 1 to avoid computing $\log(0)$ in log likelihood. The parameter ε can be adjusted if needed. Based on our empiric experience, it does not notably affect output. The default value of ε in software implementation is set to 0.005 in this study. However, the decrease of ε can slightly increase the number of regulatory relations in output. For the functions, which have indefinite output for one of inputs (f_3-f_6), we use the marginal probability of the second gene to be 1 or 0 as weight function. With all notations explained above, the likelihoods corresponding to all possible six functions between two genes can be written as follows:

$$L_{\tau,f1} = \prod_t [(p_{00}{}^t/q_{02} + p_{11}{}^t/q_{12}) \cdot (1 - \varepsilon) + (p_{01}{}^t/q_{12} + p_{10}{}^t/q_{02}) \cdot \varepsilon],$$
$$L_{\tau,f2} = \prod_t [(p_{01}{}^t/q_{12} + p_{10}{}^t/q_{02}) \cdot (1 - \varepsilon) + (p_{00}{}^t/q_{02} + p_{11}{}^t/q_{12}) \cdot \varepsilon],$$
$$L_{\tau,f3} = \prod_t [p_{00}{}^t \cdot (1 - \varepsilon)/q_{02} + p_{10}{}^t + p_{11}{}^t + p_{01}{}^t \cdot \varepsilon/q_{12}], \qquad (7)$$

$$L_{\tau,f4} = \prod_t [p_{01}{}^t \cdot (1-\varepsilon)/q_{12} + p_{10}{}^t + p_{11}{}^t + p_{00}{}^t \cdot \varepsilon/q_{02}],$$
$$L_{\tau,f5} = \prod_t [p_{10}{}^t \cdot (1-\varepsilon)/q_{02} + p_{00}{}^t + p_{01}{}^t + p_{11}{}^t \cdot \varepsilon/q_{12}],$$
$$L_{\tau,f6} = \prod_t [p_{11}{}^t \cdot (1-\varepsilon)/q_{12} + p_{00}{}^t + p_{01}{}^t + p_{10}{}^t \cdot \varepsilon/q_{02}].$$

Similarly, $L_{\tau,not\ linked} = \prod_t [p_{11}{}^t + p_{00}{}^t + p_{10}{}^t + p_{01}{}^t]$.

The largest of the likelihoods $L_{f1} \ldots L_{f6}$ will suggest the function $\hat{f}$ which is the best in explaining relation between two genes for given time-delay τ. At the same time we need to find optimal time delay between genes. Thus we repeat procedure for all possible time delays and choose the one which corresponds to the largest difference in log-likelihoods of two models.

Significance of the established link is measured with p-value. Under the null hypothesis, the test statistic of $2 \cdot \log(R)$ can be approximated by the Chi-square distribution.

In summary, the algorithm of link identification can be formulated as follows:

$$(x_1, x_2, \hat{f}, \hat{\tau}) = \text{argmax}(\max_{\substack{\tau_i \\ f(x_1)}} l_{\ linked} - l_{\ not\ linked}).$$

Test 2

Figure 1(B) provides the graphical representation for two models: model M_{direct} assumes that gene x_1 regulates x_2 directly, while $M_{indirect}$ assumes that there is intermediate gene x_3 between genes x_1 and x_2. We assign τ' to be the time delay between x_1 and x_3, and τ'' to be the time delay between x_3 and x_2. After the significant pairwise dependencies found by Test 1, Test 2 will test each link $(x_1, x_2, \hat{f}_{1-2}, \hat{\tau})$ such that links of $(x_1, x_3, \hat{f}_{1-3}, \hat{\tau}')$ and $(x_3, x_2, \hat{f}_{3-2}, \hat{\tau}'')$ exist and their time delays satisfy $\tau' + \tau'' \leq \tau$.

The corresponding likelihoods of direct model M_{direct} and indirect model $M_{indirect}$ in Figure 1(B) can be expressed as next:

$$L_{direct} = \prod_t \sum_{\substack{x_1(t) \\ x_3(t+\tau') \\ x_2(t+\tau)}} \begin{aligned} &P(x_1(t)) \cdot P(x_3(t+\tau')|x_1(t)) \cdot P(x_2(t+\tau)|x_3(t+\tau'), x_1(t)) \cdot \\ &\cdot P(y_1(t)|x_1(t)) \cdot P(y_3(t+\tau')|x_3(t+\tau')) \cdot P(y_2(t+\tau)|x_2(t+\tau)), \end{aligned}$$

$$L_{indirect} = \prod_t \sum_{\substack{x_1(t) \\ x_3(t+\tau') \\ x_2(t+\tau)}} P(x_1(t)) \cdot P(x_3(t+\tau')|x_1(t)) \cdot P(x_2(t+\tau)|x_3(t+\tau')) \cdot \\ \cdot P(y_1(t)|x_1(t)) \cdot P(y_3(t+\tau')|x_3(t+\tau')) \cdot P(y_2(t+\tau)|x_2(t+\tau)).$$

However, it is unnecessary to compute all parts since we are only interested in the difference, that is, the unary function of $f(x_3)$ against the binary function of $F(x_1, x_3)$. Since the link $x_1 \rightarrow x_3$ is present in both models and it does not contribute to models differentiation, we can remove it from computation. Thus the corresponding likelihoods for two models can be written as next:

$$L_{direct} = \prod_t \sum_{\substack{x_1(t) \\ x_3(t+\tau') \\ x_2(t+\tau)}} P(x_1(t)) \cdot P(x_3(t+\tau')) \cdot P(x_2(t+\tau)|x_3(t+\tau'), x_1(t)) \cdot P(y_1(t)|x_1(t))$$

$$\cdot P(y_3(t+\tau')|x_3(t+\tau')) \cdot P(y_2(t+\tau)|x_2(t+\tau))$$

$$L_{indirect} = \prod_t \sum_{\substack{x_3(t+\tau') \\ x_2(t+\tau)}} P(x_3(t+\tau')) \cdot P(x_2(t+\tau)|x_3(t+\tau')) \cdot P(y_3(t+\tau')|x_3(t+\tau')) \cdot$$
$$P(y_2(t+\tau)|x_2(t+\tau)).$$

After applying Bayes' theorem and all reductions similar to Test 1, the likelihood of direct model M_{direct} and indirect model $M_{indirect}$ can be written as follows:

$$L_{direct} = \prod_t \sum_{\substack{x_1(t) \\ x_3(t+\tau') \\ x_2(t+\tau)}} \frac{P(x_2(t+\tau)|x_3(t+\tau'), x_1(t))}{P(x_2(t+\tau))} \cdot P(x_1(t)|y_1(t)) \cdot P(x_3(t+\tau')|y_3(t+\tau'))$$

$$\cdot P(x_2(t+\tau)|y_2(t+\tau)),$$

$$L_{indirect} = \prod_t \sum_{\substack{x_3(t+\tau') \\ x_2(t+\tau)}} \frac{P(x_2(t+\tau)|x_3(t+\tau'))}{P(x_2(t+\tau))} \cdot P(x_3(t+\tau')|y_3(t+\tau')) \cdot P(x_2(t+\tau)|y_2(t+\tau)).$$

Similarly to (6), we have

$$P(x_2(t + \tau)|x_1(t), x_3(t + \tau'))$$
$$= \begin{cases} \delta[x_2(t + \tau) = F(x_1(t), x_3(t + \tau'))], & \text{if } F(x_1(t), x_3(t + \tau')) \neq \Omega; \\ P(x_2(t + \tau)), & \text{if } F(x_1(t), x_3(t + \tau')) = \Omega; \end{cases} \Omega = \{0 \text{ or } 1\}.$$

Analogously to (7), we can write some examples of the formulas for the above mentioned likelihoods as follows:

$$L_{\tau,\tau',F1} = \prod_t \left[\begin{matrix} (p_{000}{}^t/q_{02} + p_{010}{}^t/q_{02} + p_{100}{}^t/q_{02} + p_{111}{}^t/q_{12})(1 - \varepsilon) + \\ + (p_{001}{}^t/q_{12} + p_{011}{}^t/q_{12} + p_{101}{}^t/q_{12} + p_{110}{}^t/q_{02})\varepsilon \end{matrix} \right],$$

$$L_{\tau,\tau',F11} = \prod_t \left[\begin{matrix} (p_{010}{}^t/q_{02} + p_{100}{}^t/q_{02} + p_{110}{}^t/q_{02})(1 - \varepsilon) + p_{000} + p_{001} + \\ + (p_{011}{}^t/q_{12} + p_{101}{}^t/q_{12} + p_{111}{}^t/q_{12})\varepsilon \end{matrix} \right].$$

Among all candidates, we choose such intermediate gene in model M_{direct} which can maximize likelihood of this model and therefore maximize difference between models. In the last step, we need to make sure that we choose optimal time delay τ'' between intermediate gene x_3 and target gene x_2. Thus, we select τ' such that the largest likelihood ratio $R_{x3}(\tau, \tau')$ refers to the best choice of x_3.

On the whole, the Test 2 procedure can be expressed as next:

$$(x_1, x_3, x_2, \hat{F}, \hat{\tau}, \widehat{\tau'}) = \underset{\tau'}{\operatorname{argmax}} \max \left(\max_{x_3} \max_{F(x_1,x_3)} l_{direct} - \max_{f(x_3)} l_{indirect} \right).$$

DISCUSSION

Among all the widely-used methods of GRN reverse-engineering, only Boolean networks and Boolean function networks are able to annotate relations between genes with Boolean functions. BN can be very useful in finding dependencies among genes and describing detailed dynamic of gene interactions, including steady states and attractors. However the exhaustive search of the optimal Boolean network is infeasible for the study of a large number of genes. The BFN algorithm has the computational complexity of $O(n^3)$ in the worst case when both tests are applied and the GRN is a complete directed graph. However, these two tests are conducted in sequence, not in a nest loop. This allows significant reduction of computational complexity because only a limited number of gene pairs passes Test 1 and enters Test 2.

In fact, the number of gene pairs that will require test of directness depends heavily on number of time-points and quality of measurements in dataset. When given dataset of decent quality i.e., without missing data and having sufficient number of time points i.e., more than 30, it is possible to apply Test1 only, without missing too much in reconstruction accuracy. In this case, since only pairwise relations between genes are considered, the task is reduced to linear complexity in respect to the number of genes.

As accuracy of de-novo reconstruction of real high-throughput biological systems tends to remain quite low as of today, it can be especially beneficial to apply Test1 of BFN method in ensemble with other GRN inference methods of relatively low complexity. For example, results of BFN can enrich results obtained with information theory based methods (MRNET, ARACNE and CLR) which are the most widely used in the field of whole-genome GRN reconstruction. However, when dealing with smaller number of genes, i.e., when one is interested in gene relations within specific biological pathway, it is reasonable and recommended to apply both tests.

Another feature of BFN is its ability to easily incorporate prior knowledge about regulators. If set of transcription factors and co-factors (genes that modulate TF activity) for some organism is already known, it is possible to limit the pool of possible source genes to the TFs and co-TFs only. The result of such BFN analysis can be naturally divided into groups by function and corresponding time-delay, which relate target gene to specific TF. Following gene ontology (GO) enrichment analysis of such groups can reveal potential functional role of the regulators. In our recent study [13] we apply BFN method to the time-course transcriptome datasets of mouse and rat liver to identify novel circadian regulators in mammals. The computational results correlate well with previous genome-scale and small-scale circadian studies. Moreover, 40 out of 134 novel candidate circadian regulators were confirmed by ChIP-seq experiments to have binding sites near core circadian clock genes. The proposed circadian candidate genes can provide insights for future investigation in the fields of endocrinology, psychology, and cancer research as metabolic disorders, mental disorders and cancer diseases were shown to be tightly connected to circadian rhythms.

ACKNOWLEDGMENT

The research was funded by the Ministry of Science and Technology in Taiwan, ROC, grant number 107-2118-M-009-006-MY3.

REFERENCES

[1]   Langfelder, Peter. & Horvath, Steve. (2008). "WGCNA: an R package for weighted correlation network analysis." *BMC bioinformatics*, *9*, 559. doi:10.1186/1471-2105-9-559.

[2]   Butte, Atul J. & Kohane, Isaac S. (2000). "Mutual information relevance networks: functional genomic clustering using pairwise entropy measurements." *Pacific Symposium on Biocomputing*, *5*, 415-426.

[3]   Meyer, Patrick E., Kontos, Kevin., Lafitte, Frederic. & Bontempi, Gianluca. (2007). "Information-theoretic inference of large transcriptional regulatory networks." *EURASIP J Bioinform Syst Biol*, *1*, 79879. doi:10.1155/2007/79879.

[4]   Margolin, Adam A., Nemenman, Ilya., Basso, Katia., Wiggins, Chris., Stolovitzky, Gustavo., Favera, Riccardo D. & Califano, Andrea. (2006). "ARACNE: an algorithm for the reconstruction of gene regulatory networks in a mammalian cellular context." *BMC Bioinformatics*, *7*, Suppl 1, S7. doi:10.1186/ 1471-2105-7-S1-S7.

[5]   Faith, Jeremiah J., Hayete, Boris., Thaden, Joshua T., Mogno, Ilaria., Wierzbowski, Jamey., Cottarel, Guillaume., Kasif, Simon., Collins, James J. & Gardner, Timothy S. (2007). "Large-scale mapping and validation of Escherichia coli transcriptional regulation from a compendium of expression profiles." *PLoS Biol*, *5*, e8. doi:10.1371/journal.pbio.0050008.

[6]   Simak, Maria, Yeang, Chen-Hsiang. & Lu, Henry Horng-Shing. (2017). "Exploring candidate biological functions by Boolean Function Networks for Saccharomyces cerevisiae." *PLoS One*, *12*, e0185475. doi: 10.1371/journal.pone.0185475.

[7]   Huynh-Thu, Vân Anh., Irrthum, Alexandre., Wehenkel, Louis. & Geurts, Pierre. (2010). "Inferring Regulatory Networks from Expression Data Using Tree-Based Methods." *PLoS One*, *5*, e12776. doi:10.1371/journal.pone.0012776.

[8]   Opgen-Rhein, Rainer. & Strimmer, Korbinian. (2007). "From correlation to causation networks: a simple approximate learning algorithm and its application to high-dimensional plant gene expression data." *BMC. Syst. Biol.*, *1*, 37. doi:10.1186/1752-0509-1-37.

[9]   Müssel, Christoph., Hopfensitz, Martin. & Kestler, Hans A. (2010). "BoolNet - an R package for generation, reconstruction and analysis of Boolean networks." *Bioinformatics*, *26*, 1378-1380. doi: 10.1093/bioinformatics/btq124.

[10]   Lebre, Sophie. "G1DBN: A package performing dynamic Bayesian network inference. Version: 3.1.1" Published: 2013-09-05.

[11]   Spellman, Paul T., Sherlock, Gavin., Zhang, Michael Q., Iyer, Vishwanath R., Anders, Kirk., Eisen, Michael B., Brown, Patrick O., Botstein, David. & Futcher, Bruce. (1998). "Comprehensive identification of cell cycle–regulated genes of the yeast Saccharomyces cerevisiae by microarray hybridization." *Molecular Biology of the Cell*, *9*, 3273–3297. doi: 10.1091/ mbc.9.12.3273.

[12]   Cherry, J. Michael., Hong, Eurie L., Amundsen, Craig., Balakrishnan, Rama., Binkley, Gail., Chan, Esther T., Christie, Karen R., Costanzo, Maria C., Dwight, Selina S., Engel, Stacia R., Fisk, Dianna G., Hirschman, Jodi E., Hitz, Benjamin C., Karra, Kalpana., Krieger, Cynthia J., Miyasato, Stuart R., Nash, Rob S., Park, Julie., Skrzypek, Marek S., Simison, Matt., Weng, Shuai. & Wong, Edith D. (2012). "Saccharomyces Genome Database: the genomics resource of budding yeast." *Nucleic Acids Res.*, *40*, D700-5. doi: 10.1093/nar/gkr1029.

[13] Simak, Maria, Lu, Henry Horng-Shing. & Yang, Jinn-Moon. (2019). "Boolean function network analysis of time-course liver transcriptome data to reveal novel circadian transcriptional regulators in mammals." *Journal of the Chinese Medical Association.* doi: 10.1097/ JCMA.0000000000000180.

In: Boolean Logic, Expressions and Theories   ISBN: 978-1-53616-985-0
Editor: Victoria C. Carlsen   ⓒ 2020 Nova Science Publishers, Inc.

Chapter 2

STAR-FINITE LOGICS
AND THEIR ALGEBRAIZATIONS

Miklós Ferenczi[*]

Department of Algebra, Budapest University of Technology,
Budapest, Hungary

Abstract

Classical propositional logic can be regarded as a logic equipped with
the finite disjunctions $\bigvee_{i=1}^{n}$ and conjunctions $\bigwedge_{i=1}^{n}$ for every natural number
$n \geq 2$ (in addition to the negation $\neg$) – instead of the usual binary disjunction $\bigvee$ and conjunction $\bigwedge$ only. This variant of the classical propositional
logic can be extended so that instead of the naturals n, *hyperfinite* naturals
(i.e., *hypernaturals*) N are choosen, i.e., the operations $\bigvee_{i=1}^{N}$ and $\bigwedge_{i=1}^{N}$ are assumed. This extended *Boolean logic* is introduced in the paper, denoted
by $L_T^{\mathcal{L}}$, called "star-finite", or "hyperfinite" logic ("star-finite" means in
non-standard analysis an "infinite" whose behaviour is "like the finite").
The logic $L_T^{\mathcal{L}}$ is closely related to *infinitary logics*. Infinitary logics have
been investigated extensively. It is proven that the star-finite logic introduced here is even stronger, in a sense, than infinitary logics, in general.

Classical proposition logic has *Boolean algebra* as its associated concept of algebraization. It is shown that with star-finite logic the non-
standard enlargement of Boolean algebra can be associated. It is proven

---

[*]Corresponding Author's E-mail: ferenczi@math.bme.hu.

that such an extended Boolean algebra is isomorphic to a Boolean set algebra closed under the hyperfinite unions and intersections. This is remarkable because Boolean γ-algebras are not representable by Boolean γ-set algebras, if γ is an infinite cardinality. Also a new version of the classical Stone theorem is proved: Boolean algebras are isomorphic to some Boolean set algebra with a *hyperfinite* unit.

Keywords: infinitary logics, Boolean logics, star-finite, hyperfinite, enlargements of Boolean algebras

AMS Subject Classification: 03H05, 03C75, 03G05

1. INTRODUCTION

Classical infinitary propositional logics, i.e., logics with infinite conjunction and disjunction, have been investigated extensively (see [9], [4]). The importance of these logics is, among other things, that their expressive power is strong, infinite conjunctions and disjunctions replace in a sense the roles of the quantifiers. But these logics have some deficiencies, for example, compactness fails to be true if uncountable conjunctions and disjunctions are allowed.

Such *Boolean logics*, non-classical infinitary propositional logics are introduced here which are defined in a non-standard framework of non-standard analysis and the infinite conjunctions and disjunctions that they have are *hyperfinite* operations. It is shown that the properties of these logics are more fine than those of classical infinitary propositional logics, in general. For example, satisfiability and semantical consistency coincide in these logics, in contrast with some classical infinitary propositional logics. The logics introduced here are called *star-finite* (*-finite), or *hyperfinite* logics.

As is known, classical proposition logic has *Boolean algebra* as its associated concept of algebraization. It is shown that with *-finite logics the nonstandard enlargements of Boolean algebras can be associated. The non-standard enlargements, in particular, the hyperfinite enlargements of Boolean algebras are interesting also for themselves, because they have several unusual properties. It is proven that these extended Boolean algebras are closed under the *hyperfinite* sums and products, and, furthermore, they are representable, i.e., the following theorem of Stone type is valid: such an extended Boolean algebra is isomorphic to a Boolean set algebra closed under the hyperfinite unions and intersections. This is remarkable because ordinary Boolean γ-algebras are not representable by Boolean γ-set algebras, where γ is any infinite cardinal. It is proved that the foregoing enlargements of Boolean algebras are ω-compact. Also another version of the Stone theorem is proved: classical Boolean algebras are isomorphic to some Boolean set algebra with a *hyperfinite unit.*

In section 2 the non-standard-, and in particular, the hyperfinite enlargements of Boolean algebras are investigated. In section 3 the star-finite (*-finite) logics are introduced and characterized.

2. NON-STANDARD ENLARGEMENTS OF BOOLEAN ALGEBRAS

First, some concepts are introduced. We use concepts and results from non-standard analysis. Throughout the paper we work in a non-standard framework, in a suitable *superstructure* for Boolean algebras. The existence of a usual enlargement is assumed. The knowledge of basic concepts as *-transform, *-finiteness (hyperfiniteness), hypernatural, internality, etc. are assumed as prerequisites (see [10], [7], [3], [2]).

As is known $^*\mathcal{P}(X)$ is the set of the internal subsets of $^*X$ and $^*\mathcal{P}(X) \subseteq \mathcal{P}(^*X)$ holds (equality holds if X is finite). N denotes the set of natural numbers and $^*N$ denotes the enlargement of N (the set of hypernaturals). A set K in the enlargement is *-finite (hyperfinite) if there is an internal bijection between K and some initial segment ($\{0, 1, 2, \ldots Q\}$, where Q is a hypernatural) of the hypernaturals . An element b is standard if $b = {}^*a$ for some element a in the basic universe.

An internal sequence $\langle a_i : i \in Q \rangle$, where Q is a fixed hypernatural, is called a *Q-sequence.*

Next some concepts related to *Boolean algebras* are listed ([8]).

Let $\mathcal{P}_F(C)$ denote the collection of the finite subsets of C, where C is a fixed subset of a Boolean algebra $\mathcal{A}$. The concept of ultrafilter in a Boolean algebra is assumed to be known, but, nevertheless, we provide a definition:

$U \subseteq A$ is an *ultrafilter* in a Boolean algebra $\mathcal{A}$ if
(i) $1 \in U$
(ii) $x, y \in U$ implies $x \cdot y \in U$
(iii) $x \in U$ and $y \in A$ imply $x + y \in U$

Miklós Ferenczi

(iv) $x \in U$ or $-x \in U$ holds for every $x \in A$.

A Boolean algebra B is ω-*compact* if for an arbitrary *countable* subset S of B such that the finite product property holds in S (i.e., the product of any finitely-many elements of S is non-zero), the product of the elements of S exists and is non-zero.

The infinite sums and products of a set of elements in a Boolean algebra will be called also suprema and infima of the set.

A Boolean algebra A is called *internal* if the basic set of A and the Boolean operations are internal ones. In an internal Boolean algebra an ultrafilter U is *internal* if the set U is internal.

A Boolean algebra B is said to be *hyperfinitely-closed* if the Boolean suprema $\sum_{a \in Q} a$ and infima $\prod_{a \in Q} a$ exist in B for any hyperfinite $Q \subseteq B$. A Boolean *set algebra* B is said to be *hyperfinitely-closed as set algebra* if the unions $\bigcup_{a \in Q} a$ and intersections $\bigcap_{a \in Q} a$ exist in B for any hyperfinite $Q \subseteq B$, where $\bigcup$ and $\bigcap$ mean the union and the intersection of sets, respectively. We remark that the hyperfinitely closedness is an unusual property in the theory of Boolean algebras.

The Boolean algebras A and B in the enlargement are isomorphic if there is an internal, one-to-one mapping between them, preserving the Boolean operations.

If A is a ordinary Boolean set algebra with unit X, then A is said *minimally-generated* by a set $Y \subset A$, if A is the minimal Boolean set algebra with unit X including the set Y. If B is an internal Boolean set algebra with the unit U, then B is said *minimally-generated by an internal set* $V \subset B$ if B is the minimal one among the internal Boolean set algebras with unit U, including the set V. The definition of the *minimally-generated hyperfinite* Boolean set algebra by a hyperfinite set is analogous.

Theorem 1. *Let A be a Boolean algebra and* $^*A$ *be the enlargement of A. Then*

(i) $^*A$ *is hyperfinitely-closed*

(ii) $^*A$ *has such a Boolean set algebra representation $\tilde{A}$ that $\tilde{A}$ is hyperfinitely-closed as set algebra, the hyperfinite infima and suprema in* $^*A$ *are preserved at the canonical isomorphism and the members of $\tilde{A}$ are internal sets*

(iii) $\tilde{\mathcal{A}}$ *is ω-compact*

(iv) if $\mathcal{A}$ is a Boolean set algebra with unit X and $\mathcal{A}$ is minimally-generated by a set $Y \subset A$, then $^*\mathcal{A}$ is minimally-generated by the set $^*Y$.

Proof.

(i) It follows from the assumptions that $^*\mathcal{A}$ is the *-transform of $\mathcal{A}$. The detailed proof is as follows.

$\sum_{a \in F} a$ is a function defined on $\mathcal{P}_F(\mathcal{A})$ (finite subsets of A) mapping into A. $^*\mathcal{P}_F(A)$ is exactly the hyperfinite subsets of $^*\mathcal{A}$. The *-extension $^*\sum$ is defined on $^*\mathcal{P}_F(A)$ and maps into $^*A$.

Let us assume that Q is hyperfinite and $Q \subseteq {^*A}$. We state that

$$^*\sum_{a \in Q} a = \sum_{a \in Q} a \tag{1}$$

where on the right hand side Σ denotes the usual Boolean supremum in $^*\mathcal{A}$.

The finite Boolean sums $\sum_{a \in Z} a$ as operations have the following characterization for finite subsets $K \subseteq A$:

$$\forall Z \in \mathcal{P}_F(A) \forall u \in Z(u \leq \sum_{a \in Z} a \wedge \exists v \in A \forall t \in Z(t \leq v \rightarrow \sum_{a \in Z} a \leq v)).$$

The $*$-transform of this formula is:

$$\forall Z \in {^*\mathcal{P}_F(A)} \forall u \in Z(u \; ^* \!\!\leq \; ^*\!\!\sum_{a \in Z} a \wedge \exists v \in {^*A} \forall t \in Z(t \; ^* \!\!\leq v \rightarrow \; ^*\!\!\sum_{a \in Z} a$$
$$^* \!\!\leq v)).$$

It says that for arbitrary hyperfinite Z in $^*A$ ($Z \in {^*\mathcal{P}_F(A)}$) $^*\!\!\sum_{a \in Q} a$ is the usual Boolean supremum of the elements in Z.

The proof of (1) for the infimum $\prod_{a \in Q} a$ is similar.

(ii) Let the desired definition of the representation $\tilde{\mathcal{A}}$ of $^*\mathcal{A}$ be the following one:

Let the unit H be the set $\{U : U$ is an *internal* ultrafilter in $^*\mathcal{A}\}$. Let us consider the following mapping f from $^*\mathcal{A}$ into $\mathcal{P}(H)$:

$$fa = \{U : a \in U, U \text{ is internal ultrafilter in } {^*\mathcal{A}}\} .$$

First, notice that fa is also *internal* in $\mathcal{P}(H)$.

Let us consider the formalised definition of fa :

$$\{U : U \in {}^*\mathcal{P}(A) \wedge^* JU \wedge a \in U \wedge a \in {}^*A\} \qquad (2)$$

where J denotes the ultrafilter property of U :

$$1 \in U \wedge \forall u \in U \forall v \in A(u + v \in U \wedge (v \in U \to u \cdot v \in U)). \qquad (3)$$

By the internal definition principle, f and fa are really internal because the constants are internal in the formal definition (2).

We check that f is an embedding of ${}^*\mathcal{A}$ into the power set algebra $\mathcal{P}(H)$.

f is injective because every element in ${}^*\mathcal{A}$ is included in some internal ultrafilter U. Namely by the consequence of the ultrafilter theorem

$$\forall x \in A(x \neq 0 \to \exists U \in \mathcal{P}(A)(JU \wedge x \in U)) \qquad (4)$$

holds, where J means the ultrafilter property (3). Applying the transfer principle we obtain the property

$$\forall x \in {}^*A(x \neq 0 \to \exists U \in {}^* \mathcal{P}(A)({}^*JU \wedge x \in U))$$

which means that every element in ${}^*\mathcal{A}$ is included in some internal ultrafilter U.

To prove that f preserves the finite Boolean operations is routine. As a consequence, f is really an embedding of ${}^*\mathcal{A}$ into $\mathcal{P}(H)$.

Let us notice that the internal ultrafilters in ${}^*\mathcal{A}$ are closed with respect to the hyperfinite suprema and infima. Namely the ultrafilter property (3) can be modified in such a way that it should mean the property

$$1 \in U \wedge \forall u \in U \, \forall v \in A(u + v \in U \wedge \forall W \in \mathcal{P}_F(U)(\prod_{a \in W} a \in U \wedge \sum_{a \in W} a \in U)).$$

At the transfer $*$, $\mathcal{P}_F(U)$ goes to the set ${}^*\mathcal{P}_F(U)$, i.e., to the set of the hyperfinite subsets of the internal ultrafilter U, furthermore $\prod$ and $\sum$ go to the hyperfinite ${}^*\prod$ and ${}^*\sum$. These latter extensions are Boolean infima and suprema (see (1)).

Then we show that the preservation of the hyperfinite Boolean operations at f is guaranteed by the above preservation property of the internal ultrafilters. For example, we check that f preserves the hyperfinite infima, i.e.,

$$f\left({}^* \prod_{a \in D} a \right) = \bigcap_{a \in D} fa \tag{5}$$

holds, where D is a hyperfinite set in $^*A$.

The part $\supseteq$ in (5). Let $C \in \bigcap_{a \in D} fa$. Then, $C \in fa$ for every $a \in D$, i.e., $a \in C$ for every $a \in D$. The internal ultrafilter C is closed under the hyperfinite infima, thus $^* \prod_{a \in Q} a \in C$. This means that $C \in f\left({}^* \prod_{a \in Q} a \right)$.

The part $\subseteq$ in (5) follows from the definition of the operation intersection.

The proof of the preservation of the hyperfinite suprema is completely analogous.

(5) implies that $\tilde{A}$ is hyperfinitely-closed as set algebra, since, if $D' \subseteq \tilde{A}$ and D' is hyperfinite, then it is obvious that $f^{-1}D'$ is also hyperfinite subset of $^*A$. Applying (5) for $D = f^{-1}D'$ we obtain the hyperfinitely-closedness of $\tilde{A}$.

(iii) By (ii), the members of $\tilde{A}$ are internal. The proposition follows from the known saturation property of the internal sets (see [10], Ch.4).

(iv) if Y minimally generates $\mathcal{A}$, then the following property holds:
$$Y \subseteq A \wedge \forall Z \in \mathcal{P}(\mathcal{P}(X))[Y \subseteq Z \wedge \text{Booleanset}(Z) \to Y \subseteq Z]$$
where $\text{Booleanset}(Z)$ denotes the formula
$$\forall U \forall V \in Z \, (U \cap V \in Z \wedge U \cup V \in Z \wedge \sim U \in Z).$$

At the $*$-transform this formula goes to the formula
$$^*Y \subseteq {}^*A \wedge \forall Z \in {}^*\mathcal{P}(\mathcal{P}(X))[{}^*Y \subseteq Z \wedge {}^*\text{Booleanset}(Z) \to {}^*Y \subseteq Z].$$

This means the following: among the internal Boolean set algebras with unit $^*X$, containing the set $^*Y$, the minimal one is $^*A$, i.e., $^*Y$ minimally-generates $^*A$.

$\square$

Remarks.

- Part (ii) can also be proved in such a way that we set out from a usual set
  representation $\mathcal{A}'$ of the Boolean algebra $\mathcal{A}$ and we take the enlargements
  of $\mathcal{A}$ and $\mathcal{A}'$, simultaneously.

  In this case let $\widetilde{\mathcal{A}}$ be $^*(\mathcal{A}')$. We show that the hyperfinite infima and
  suprema in the enlargement are exactly the set operations intersection and
  union. As regards the union for example, this follows from the definition
  of the union. Let us consider the union axiom for finite unions in $\mathcal{A}'$:

  $$\forall x \in \mathcal{P}_F(A')\exists y \in A'\forall u(u \in y \leftrightarrow \exists v(u \in v \wedge v \in x)).$$
  The $*$-transform of this formula is:

  $$\forall x \in {}^*\mathcal{P}_F(A')\exists y \in {}^*(A')\forall u(u \in y \leftrightarrow \exists v(u \in v \wedge v \in x)).$$
  It says that for a hyperfinite set x $(x \in {}^*\mathcal{P}_F(A'))$ the union of the members
  of x exists in $^*(\mathcal{A}')$.

  The members of $\widetilde{\mathcal{A}}$ are internal, by definition. Composing the inverse of
  the transformation * defined on $\mathcal{A}$, the ordinary canonical isomorphism
  and the embedding * of $\mathcal{A}'$, we get the canonical isomorphism from $^*\mathcal{A}$
  onto $\widetilde{\mathcal{A}}$. It obviously preserves the hyperfinite operations.

- Notice that while by (ii) the algebra $\widetilde{\mathcal{A}}$ includes the hyperfinite intersections and unions (these are uncountable operations if they are infinite
  ones), $\widetilde{\mathcal{A}}$ does not include the countable intersections and unions, only
  finite ones (because the saturated property of internal sets).

- (iii) can be generalized from ω-compactness to $\varkappa$-compactness, where
  $\varkappa$ is arbitrary fixed infinite cardinal, because the so-called $\varkappa$-saturation
  property of the internal sets (see [10], Ch.4).

Now we return to another extension of Boolean algebras. In [7] 19.4 it is
proved that the following proposition holds:

For every Boolean algebra $\mathcal{A}$ there is a hyperfinite, atomic Boolean algebra,
denoted by $\mathcal{A}^+$, such that $\mathcal{A} \subset \mathcal{A}^+ \subset {}^*\mathcal{A}$ holds.   (6)

For the sake of discussion below, let us remind the reader of the main points of the proof.

We start with a well-known definition. Let a binary relation R, defined on a set $U \times V$, be an object in the basic universe. R is *concurrent* if for any finite subset $\{a_1, a_2 \ldots a_n\}$ of U there exists an element $a \in V$ such that $a_i R a$ if $1 \leq i \leq n$. Then, the *enlargement* contains such an element $b \in^* V$ that $^*c(^*R)b$ for every element $c \in U$.

Now, the sketch of the proof in [7] 19.4 is as follows —

Let A be the basic set of the Boolean algebra $\mathcal{A}$. Let the definition of a relation R, defined on the set $A \times \mathcal{P}_F(A)$, be the following:

$$(a, D) \in R \text{ if and only if } a \in \mathcal{D}, \text{ where } \mathcal{D} \text{ is an atomic finite Boolean}$$

subalgebra of $\mathcal{A}$.   (7)

R is concurrent because if $a_1, a_2 \ldots a_n \in A$, then $\mathcal{D}$ can be chosen as the Boolean algebra generated by $\{a_1, a_2 \ldots a_n\}$. This latter algebra is known to be finite and atomic.

Let us consider the enlargement $^*\mathcal{A}$. Then, $^*R \subseteq {}^*\mathcal{A} \times {}^*\mathcal{P}_F(A)$. By the concurrency of R there exists such an element $A^+ \in {}^*\mathcal{P}_F(A)$ that

$$^*a \, ^*R \, A^+ \qquad (8)$$

holds for every $a \in A$, i.e., $^*a \in {}^+A$ for every $a \in A$.

On the one hand, $A^+ \in {}^*\mathcal{P}_F(A)$ means that A^+ is a hyperfinite subset of $^*A$. On the other hand, by the transfer principle and by the definition of R, $\mathcal{A}^+$ is a Boolean subalgebra of $^*\mathcal{A}$ and atomic, because these properties can be formalised in the language of Boolean algebras. For example, the formalization of the property "atomic" is:

$$\forall x \in A(x = 0 \rightarrow \exists a \in A(a = 0 \wedge \forall y \in A(y \leq a \rightarrow y = 0 \vee y = a) \wedge a \leq x)).$$

Thus, $\mathcal{A}$ can indeed be embedded into the hyperfinite and atomic Boolean algebra $\mathcal{A}^+$ and $\mathcal{A}^+ \subset {}^*\mathcal{A}$.

An obvious concequence is that the *hyperfinite Boole algebras are atomic*.

The following theorem is an improvement and generalization of proposition (6). The point of the statement is that the proposition of Theorem 1 remains true for the algebra $\mathcal{A}^+$, under little modifications.

Theorem 2. *Let $\mathcal{A}^+$ be the Boolean algebra in (6). Then*

(i) $\mathcal{A}^+$ is hyperfinitely-closed

(ii) $\mathcal{A}^+$ has a Boolean set algebra representation $\widetilde{\mathcal{A}^+}$ with a hyperfinite unit, which is hyperfinitely-closed as set algebra, the canonical isomorphism preserves the hyperfinite infima and suprema, furthermore the members of $\widetilde{\mathcal{A}^+}$ are hyperfinite sets

(iii) $\widetilde{\mathcal{A}^+}$ is ω-compact

(iv) if $\mathcal{A}$ is a Boolean set algebra with unit X and $\mathcal{A}$ is minimally-generated by a set $Y \subseteq A$, then $\mathcal{A}^+$ is minimally-generated by the set $^*Y$.

Proof. We refer to the above sketched proof of (6).

(i) Let us supplement the definition (7) of the relation R with the property that $\mathcal{D}$ is closed under the finite products and sums, i.e., by the property
$$\forall W \in \mathcal{P}_F(D)(\prod_{a \in W} a \in D \wedge \sum_{a \in W} a \in D).$$ At the enlargement, this property is transformed into the property that the hyperfinite subsets of $\mathcal{A}^+$ are closed under the extensions $^*\prod$ and $^*\sum$.

Modifying the definition of R, by the definition of finite supremum and infimum (similarly to the proof of (1)) we can prove that the extensions $^*\prod$ and $^*\sum$ are Boolean infima and suprema in $\mathcal{A}^+$.

(ii) It is known that every atomic Boolean algebra is isomorphic to a Boolean set algebra whose unit is the set of atoms (see [8]). Let us denote this set algebra by $\widetilde{\mathcal{A}^+}$.

As is known, the foregoing canonical, standard isomorphism is a complete isomorphism, i.e., it preserves all the infima and suprema in $\mathcal{A}^+$. This implies that $\widetilde{\mathcal{A}^+}$ is hyperfinitely-closed as Boolean set algebra. At the canonical isomorphism, with an element of $\mathcal{A}^+$ the set of atoms included in the foregoing element is associated.

The set

$$\{a : a \in A^+ \wedge a \neq 0 \wedge \forall y \in A^+ (y \leq a \rightarrow y = 0 \vee y = a)\}.$$

of atoms of $\mathcal{A}^+$ is internal, because the internal definition principle. This internal subset of the hyperfinite $\mathcal{A}^+$ is also hyperfinite, thus the unit of $\widetilde{\mathcal{A}^+}$ is really hyperfinite. Furthermore, the elements of $\mathcal{A}^+$ are also internal, because $\mathcal{A}^+ \subset {}^*\mathcal{A}$. Then, using the internality of the canonical isomorphism, we obtain that the members of $\widetilde{\mathcal{A}^+}$ are internal and also hyperfinite ones.

(iii) See the proof of Theorem 1 (iii).

(iv) We can supplement the definition (7) of the relation R in such a way that a fixed subset G of $\mathcal{D}$ minimally-generates $\mathcal{D}$, i.e., $\mathcal{D}$ is the minimal Boolean algebra among the finite Boolean subalgebras of $\mathcal{A}$, including $\dot{G}$. This property can be formalized in an analogous way as it is done in the proof of Theorem 1 (iv). Applying the transfer * for R we obtain that $^*G$ minimally-generates $\mathcal{A}^+$, i.e., $\mathcal{A}^+$ is the minimal hyperfinite Boolean algebra among the hyperfinite Boolean subalgebras of $^*\mathcal{A}$ including $^*G$.

This argument can be repeated for the case when $\mathcal{A}$ is a Boolean *set algebra* with unit X, minimally-generated by a set $Y \subset A$.

□

The following version of the Stone theorem concerning the representation of Boole algebras follows:

Corollary. *Every Boolean algebra is isomorphic to a Boolean set algebra with a hyperfinite unit.*

This follows from (ii), because (ii) implies that $\mathcal{A}$ can be embedded into the set algebra $\widetilde{\mathcal{A}^+}$ *having a hyperfinite unit.*

In a forthcoming paper we are going to generalize the results in this paragraph from Boolean algebras to cylindric algebras (see [1], [6]). In this generalization the Henkin type semantics seem to play an essential role ([5]).

3. STAR-FINITE (*-FINITE) LOGICS

In this section, besides the Boolean algebra, a classical propositional logic equipped with finite disjunctions and conjunctions as elementary operations together with the usual propositional semantics are assumed in the basic universe. We are going to specify this logic later in the paper.

Next, a special *Boolean logic*, the *-finite logic (a kind of infinitary propositional logic) is defined in the enlargement. First its language and semantics are
defined.

The abc of the language $\mathcal{L}$.

- The logical constants: for every $n \in {}^*N$, i.e., for every hypernatural, the
  symbols $\bigvee\limits_n$, $\bigwedge\limits_n$, furthermore $\neg$.

- The set of the propositional symbols: the set ${}^* \{B_j : j \in N\}$ denoted
  by A, where B_j are arbitrary propositional symbols. That is, A is a *-
  transform of a set of ordinary propositional symbols, so A is standard.

The concept of *formula* α (i.e., a well-formed formula α) in $\mathcal{L}$.
(i) The members of A are formulas.
(ii) If $\langle \alpha_i \rangle_{i \in H}$ is a H-sequence of formulas, where H is a fixed hypernatural,
then $\bigwedge\limits_H \alpha_i$, $\bigvee\limits_H \alpha_i$ are formulas, and $\neg \alpha_i$ is a formula if $i \in H$.
Formulas of $\mathcal{L}$ are obtained by applying hyperfinitely-times, say M-times,
the rules (i) and (ii), where M is a hypernatural.

The *rank of a formula* is the hypernatural M occuring in the definition of
the formula. If M is finite, then the formula is said to be a formula of finite
rank.

An *interpretation function* t for the language $\mathcal{L}$ is an internal function mapping from the set of the formulas in $\mathcal{L}$ into the Boolean algebra $\mathcal{B}$ of two elements 0 and 1, having the property (p') below:

Property (p'): if $\langle \alpha_i \rangle_{i \in H}$ is a H-sequence of formulas in $\mathcal{L}$, then

$$ t(\bigwedge\limits_H \alpha_i) = \min_{1 \leq i \leq H} \{t(\alpha_i)\}, t(\bigvee\limits_H \alpha_i) = \max_{1 \leq i \leq H} \{t(\alpha_i)\} \text{ and } t(\neg \alpha_i) = -t(\alpha_i). $$

The restriction of t to A is called *elementary interpretation*. Let us fix a
standard set T of the elementary interpretations. T is said to be the set of the
possible elementary interpretations.

The *interpretation set* (or, truth set) $|\alpha|$ of a formula α is defined as the set
$\{t : t(\alpha) = 1, t \in T\}$.

It is easy to check by formula recursion that the interpretation sets have the
following properties:

(i) if the interpretation sets $|\alpha|$ have been defined, then the interpretation set of $|\neg\alpha|$ is $T \sim |\alpha|$.

(ii) if the interpretation sets $|\alpha_i|$ of α_i have been defined for a Q-sequence $\langle \alpha_i : i \in Q \rangle$, where Q is a hypernatural, then the interpretation sets of the formulas $\bigvee_Q \alpha_i$ and $\bigwedge_Q \alpha_i$ are the sets $\bigcup_Q |\alpha_i|$ and $\bigcap_Q |\alpha_i|$, respectively. $\qquad$ (9)

The logic defined by the above semantics is called *star-finite* (**-finite*, or *hyperfinite) logic* and denoted by $L_T^{\mathcal{L}}$. Let $Lf_T^{\mathcal{L}}$ denote the restriction of $L_T^{\mathcal{L}}$, where the ranks of the formulas are *finite*. $Lf_T^{\mathcal{L}}$ is called *star-finite* (**-finite*, or *hyperfinite) logic with finite ranks*.

The concept of interpretation function implies the definitions of the main concepts of semantics. Thus, it implies the concept of satisfiability, or the logical equivalency of two formulas:

- α and β are *logical equivalent* if their interpretations t give the same value for every $t \in T$.

- A formula α is satisfiable if $t(\alpha) = 1$ for some $t \in T$.

Similarly to classical logic, $\Sigma \models \beta$ is equivalent to the unsatisfiability of the formula set $\Sigma \cup \{\neg\beta\}$.

If a classical propositional logic is modified in such a way that the usual (binary) disjunctions and conjunctions $\vee$ and $\wedge$ are replaced for every $n \in N$ by the n-ary disjunctions and conjunctions $\bigvee_n$ and $\bigwedge_n$ (i.e., by infinitely-many operations), then the logic obtained is called an ILC logic (propositional logic with Infinitely-many Logical Constants).

The following holds:

Theorem 3. *A *-finite logic is the *-transform of some ILC propositional logic and, conversely, the *-transform of an ILC propositional logic is a *-finite logic.*

Proof. Let us have an *ILC* propositional logic. As regards its *language* $\mathcal{V}$, the symbols of the n-ary disjunctions $\bigvee_n$ and conjunctions $\bigwedge_n$ are assumed, where

n runs over the natural numbers being ≥ 2 and $\mathcal{V}$ contains the unary operation symbol negation $\neg$. Furthermore, $\mathcal{V}$ includes the set $\{B_j : j \in N\}$ of propositional symbols, where N is the set of natural numbers.

    The concept of formula, i.e., that of well-formed formula (wff), is the usual:

    (i) The propositional symbols are formulas.

    (ii) If $\alpha_1, \alpha_2, \ldots \alpha_n$ is a finite sequence of formulas, where n is a fixed natural number, then $\bigwedge_n \alpha_i, \bigvee_n \alpha_i$ are formulas, and $\neg \alpha_i$ $(i \leq n)$ is a formula.

Formulas are obtained by applying finitely-many times the rules (i) and (ii). Let W denote the set of the formulas.

The semantics of this propositional logic is the usual:

An interpretation function s is a mapping from the set W of the formulas into the Boolean algebra $\mathcal{B}$ of two elements 0 and 1, having the property (p) below:

Property (p): if $\alpha_1, \alpha_2, \ldots \alpha_n$ are formulas, then

$$s(\bigwedge_n \alpha_i) = \min_{1 \leq i \leq n} \{s(\alpha_i)\}, s(\bigvee_n \alpha_i) = \max_{1 \leq i \leq n} \{s(\alpha_i)\} \text{ and } s(\neg \alpha_i) = -s(\alpha_i).$$

    Let S denote a fixed non-empty set (possible interpretations) of the interpretation functions . The *truth set* $|\alpha|$ of a formula α is the set $\{s \in S : s(\alpha) = 1\}$.

    Let $F_S^{\mathcal{V}}$ denote the ILC propositional logic introduced just before. Let $F_S^{\mathcal{V}}$ be denoted in the basic universe.

    It is easy to see that, by definition, the set of formulas in the language $\mathcal{L}$ of the logic $L_T^{\mathcal{L}}$ is $^*W$, where W is the set of the formulas in $\mathcal{V}$.

    The concept of interpretation function for the logic $L_T^{\mathcal{L}}$ stems from the facts that the Boolean algebra $\mathcal{B}$ of two elements remains unchanged at the *-transform and the set S of interpretation functions goes into a set $^*S$ $(= T)$ of internal functions. The property (p) is inherited at the *-transform from the finite set of formulas to the hyperfinite set of formulas, i.e., to the H-sequences of formulas; this is why (p') is true. Furthermore, $^* \min = \min$ and $^* \max = \max$.

    Both directions of the theorem follow from these considerations.   □

    The definition by *-transform implies that *-finite logics have many analogous properties in common with classical propositional logic. Many algebraic properties are transfered by *-transform: associativity, commutativity, distributivity of the conjunctions and disjunctions.

For example, the formulation of the normal form theorem for the logic $L_T^{\mathcal{L}}$ is the following:

Every formula of $L_T^{\mathcal{L}}$ is logical equivalent to a hyperfinite disjunction of certain hyperfinite conjunctions of literals, i.e., equivalent to a formula of the form

$$\bigvee_{i \leq Q} (\bigwedge_{j \leq K_i} L_{ij}) \tag{10}$$

where Q, K_i are hypernaturals, L_{ij} are literals and the members of the disjunctions and conjunctions in (10) constitute Q-, and K_i-sequences (i.e., internal sequences).

Namely, by the well-known theorem of classical propositional logic, for every formula α in W, there is a formula β being a finite disjunction of certain finite conjunctions of literals, logically equivalent to α, i.e., $s(\alpha) = s(\beta)$ for every s. Considering the *-transform we get that for every formula α in $^*W$ there is a formula β being a hyperfinite disjunction of hyperfinite conjunctions of literals, logically equivalent to α, i.e., $t(\alpha) = t(\beta)$ for every t. The finite sequences of the members of the finite conjunctions and disjunctions go into internal hyperfinite sequences at the *-transform.

With the infinary logic L_{ω_1} (including the operations $\bigvee_{i=1}^{\infty}$ and $\bigwedge_{i=1}^{\infty}$) the logic $Lf_T^{\mathcal{L}}$ can be associated rather than $L_T^{\mathcal{L}}$ because the ranks of the formulas are finite in both logics. There are many similar properties of L_{ω_1} and $Lf_T^{\mathcal{L}}$. Next, such a property of $Lf_T^{\mathcal{L}}$ is presented which makes a difference between these two logics.

An important semantical concept in infinary logics is the semantical consistency (see [9]). A set Λ of formulas is *semantically consistent* if every *finite* subset of Λ is satisfiable.

Recall that the equivalence of satisfiability and semantical consistency *fails* to be true for the infinary logic L_{ω_1} for infinite countable formula sets (see [9]). But, for *-finite logic the equivalence holds.

Let Λ be a set of formulas of the logics $Lf_T^{\mathcal{L}}$ or $L_T^{\mathcal{L}}$.

Theorem 4. *For any countable set Λ of formulas, Λ is satisfiable if and only if Λ is semantically consistent.*

Proof. The satisfiability of Λ implies the semantical consistency of Λ, this is trivial. We check the other direction. First, using the definition of formula in $\mathcal{L}$, we prove that the truth sets $\{t \in {}^*S : t(\alpha) = 1\}$ of formulas α are internal sets.

Let us consider the truth set of a propositional symbol C, i.e., the set $\{t \in {}^*S : t(C) = 1, C \in A\}$, where A denotes the set ${}^*\{B_j : j \in N\}$. The internal definition principle (see [10] 3.3) implies that this set is internal.

Then we use the properties of interpretations in (9). Assume that the truth sets of the formulas α and β are internal sets. By definition, these are subsets of the internal set ${}^*S$. As is known, the complement, the union and the intersection of two internal sets are also internal. And, if the truth sets of the formulas form a Q-sequence $\langle \alpha_i : i \in Q \rangle$ (i.e., form an internal sequence), then the truth sets of the formulas $\bigvee_Q \alpha_i$ and $\bigwedge_Q \alpha_i$ are also internal sets, because these are internal unions and intersections of internal sets (see [10] 3.3).

If α is in $Lf_T^{\mathcal{L}}$, i.e., α can be composed in finitely-many steps, then ordinary formula induction applies to prove that the truth sets are internal sets.

If α is in $L_T^{\mathcal{L}}$ then *internal formula induction* applies ([7] Ch.11.3) rather than ordinary formula induction.

To complete the proof, we use that every enlargement has the "countable compactness" property, i.e., the property that every countable collection of internal sets having the finite intersection property has a non-empty intersection ([10] Ch.4).   □

Remarks.

- The theorem can be generalized from countable cardinality to cardinalities less than the fixed cardinal number κ, using so-called κ-*saturated* enlargements (see [10], Ch.4).

- An easy consequence is that the logic $L_T^{\mathcal{L}}$ is ω-compact, i.e., if for a countable set Σ of formulas any finite subset of Σ is satisfiable, then Σ is also satisfiable.

Theorem 5. *The following propositions hold:*

(i) *The interpretations sets of $L_T^{\mathcal{L}}$ form a hyperfinitely-closed, standard Boolean set algebra (this algebra is called the model algebra associated with the language $\mathcal{L}$ and the interpretation set T, denoted by $\mathcal{A}_T^{\mathcal{L}}$).*

(ii) *If $\mathcal{B}$ is a hyperfinitely-closed standard Boolean set algebra, then $\mathcal{B}$ can be regarded as a model algebra $\mathcal{A}_T^{\mathcal{L}}$ for suitable $\mathcal{L}$ and T.*

(iii) *$\mathcal{A}_T^{\mathcal{L}}$ is ω-compact.*

(iv) *The model algebra $\mathcal{A}_T^{\mathcal{L}}$ is a *-transform of the model algebra $\mathcal{A}_S^{\mathcal{V}}$ corresponding to the logic $F_S^{\mathcal{V}}$.*

Proof. (i) The interpretation sets form an ordinary Boolean subalgebra of $\mathcal{P}(T)$ because the collection of the interpretation sets is closed under the operations union, intersection and complement. The proof of Theorem 4 implies that this Boolean subalgebra is closed also under the hyperfinite unions and intersections and the members of the Boolean algebra are internal. Part (iv) will imply that $\mathcal{A}_T^{\mathcal{L}}$ is standard.

(ii) $\mathcal{B}$ is standard, so its elements are internal. Let us choose a fixed standard generator system G in $\mathcal{B}$. Such a generator system exists, because $\mathcal{B}$ is standard. For every element g of G let us introduce a symbol R_g in the language and let the elementary interpretation $|R_g|$ be g. Let T be the set G. Let us consider the model algebra $\mathcal{A}_T^{\mathcal{L}}$. The definition of the generator system G implies that $\mathcal{B} \subseteq \mathcal{A}_T^{\mathcal{L}}$. The properties of the interpretation sets in (9) and the hyperfinitely-closedness of $\mathcal{B}$ imply that $\mathcal{B} \supseteq \mathcal{A}_T^{\mathcal{L}}$, i.e., $\mathcal{B}$ and $\mathcal{A}_T^{\mathcal{L}}$ coincide.

(iii) See the proof of Theorem 1 (iii).

(iv) As is known, $\mathcal{A}_S^{\mathcal{V}}$ is minimally-generated by the sets $\{|B_j|\}_{j \in N}$, where $\{B_j\}_{j \in N}$ is the set of the propositional symbols in the language $\mathcal{V}$. By Theorem 1 (iv) the enlargement of $\mathcal{A}_S^{\mathcal{V}}$ is minimally-generated by the sets in $^*(\{|B_j|\}_{j \in N})$. But, by Theorem 3 the logic $L_T^{\mathcal{L}}$ is a *-transform of the logic $F_S^{\mathcal{V}}$. So $\mathcal{A}_T^{\mathcal{L}}$ is minimally-generated by the set $^*(\{|B_j|\}_{j \in N})$. Thus $\mathcal{A}_T^{\mathcal{L}}$ coincides with the enlargement of $\mathcal{A}_S^{\mathcal{V}}$, i.e., $\mathcal{A}_T^{\mathcal{L}}$ is just the enlargement of $\mathcal{A}_S^{\mathcal{V}}$. $\square$

Remarks.

- Theorem 3, (i) and (ii) imply that the class of the model algebras associated with star-finite logics coincide with the class of hyperfinitely-closed standard Boolean set algebras.

- We obtain a particular class of logics from star-finite logics when, in addition, the set A of the propositional symbols, the set T of interpretations and the elementary interpretation sets are assumed hyperfinite. With this kind of logic the hyperfinite Boolean algebra $\mathcal{A}^+$ in Theorem 2 can be associated as algebraization.

REFERENCES

[1] Andréka, H., Ferenczi, M. and Németi, I. (2012). *Cylindric-like Algebras and Algebraic Logic*. Bolyai Society Mathematical Studies, Springer.

[2] Anderson, R. M. (1982). Star-finite representations of measure spaces. *Trans. Amer. Math. Soc.*, 271, 667-687.

[3] Andreev, P. V., and Gordon, E. I. (2006). A theory of hyperfinite sets. *Annals of Pure and Applied Logic*, 143, 3-19.

[4] Bairwise, J. (1969). Infinitary logic and admissible sets. *Journal of Symbolic Logic*, 34, 2, 226-252.

[5] Ferenczi, M. (2012). The polyadic generalization of the Boolean axiomatization of fields of sets. *Trans. Amer. Math. Soc.*, 364, 867-886.

[6] Ferenczi, M. (2018). Cylindric algebras and finite polyadic algebras. *Algebra Universalis*, 79:(60).

[7] Goldblatt, R. (1998). *Lectures on the Hyperreals: An Introduction to Nonstandard Analysis*. Graduate Texts in Mathematics 188, Springer.

[8] Givant, S., and Halmos, P. (2009). *Introduction to Boolean Algebras*. Springer.

[9] Karp, C. R. (1964). *Languages with Expressions of Infinite Length*. Amsterdam: North-Holland.

[10] Väth, M. (2007). *Non-standard Analysis*. Birkhäuser.

In: Boolean Logic, Expressions and Theories   ISBN: 978-1-53616-985-0
Editor: Victoria C. Carlsen   © 2020 Nova Science Publishers, Inc.

Chapter 3

GENERALIZED BOOLEAN FUNCTIONS AND THEIR APPLICATIONS TO COMMUNICATIONS

Fanxin Zeng*, Yue Zeng, Guojun Li,†
Lisheng Zhang and Changrong Ye
Lab. of Beyond LOS Reliable Information Transmission
Chongqing University of Posts and Telecommunications
Chongqing, China

Abstract

A generalized Boolean function (GBF) is a mapping from direct product Z_2^m to set Z_H, where $Z_2 = \{0, 1\}$, $Z_H = \{0, 1, 2, \cdots, H - 1\}$, and integers $m(\geq 1), H(\geq 2)$. GBFs can be divided into standard and non-standard ones. GBFs play fairly important roles in constructing H phase shift keying (PSK) Golay complementary sequences (GCSs). In particular, binary phase-shift keying (BPSK) or quadrature phase-shift keying (QPSK) GCSs can be resulted in for $H = 2$ or 4, respectively. For given m, standard H-PSK GCSs of length $N = 2^m$ have $m!H^{m+1}/2$ $(m, H \geq 2)$, and $(n - 2)!(n - 2)4^n$ for non-standard QPSK GCSs $(n = m + 3, m \geq 1)$. GCSs are closely associated with the control of peak envelope power (PEP) of transmitted signals in an orthogonal frequency-division multiplexing (OFDM) communication system. It

*Corresponding Author's E-mail: fzengx@cqu.edu.cn.
†Corresponding Author's E-mail: lgjsw@126.com.

has been proved that upper bound of peak-to-mean envelope power ratio (PMEPR) of transmitted signals in an OFDM communication system, employs BPSK or QPSK GCSs, does not excelled 2. Further, based on BPSK or QPSK GCSs referred to above, quadrature amplitude modulation (QAM) GCSs can be designed. In general, for an OFDM communication system employing resultant 4^q-QAM GCSs of length $N = 2^m$, upper bound of PEP of transmitted signals of this system can be controlled not to excel $\frac{6N(2^q-1)}{2^q+1}$.

Keywords: generalized Boolean functions, complementary sequences, quadrature amplitude modulation constellation, aperiodic autocorrelation functions, peak envelope power

1. INTRODUCTION

With rapid development of wireless communications, Golay complementary sequences (GCSs) are playing more important roles than they were so. GCSs are also known as Golay complementary sequence pair (CSP), in which each sequence of the pair is referred to as a Golay sequence. By employing sum of aperiodic autocorrelation functions (AACFs) of two Golay sequences in a Golay CSP, GCSs have a impulse-like sum. As a consequence, GCSs are substantially applied to synchronization, pilot, spreading codes, and control of peak envelope power (PEP) in communications [1], [2], [3]. Besides, GCSs are also used in large-scale integrated circuit testing, signal processing, etc.

There are lots of methods to construct GCSs, therein, the method based on generalized Boolean functions (GBFs) is fairly attractive, which results in a large number of GCSs available. An m-dimensional GBF $f(x_1, x_2, \cdots, x_m)$ is in fact a mapping from direct production $Z_2^m = \overbrace{Z_2 \times Z_2 \times \cdots \times Z_2}^{m}$ to ring Z_H, where m (≥ 1) and H (≥ 2) is positive integers, and $Z_H = \{0, 1, 2, \cdots, H-1\}$. In particular, when $H = 2$, GBFs degenerate to well-known Boolean functions. GBFs are closely associated with constructions of H-ary GCSs. For various H's, binary, quaternary, H phase-shift keying (PSK), quadrature amplitude modulation (QAM), and almost/near GCSs can be deduced from GBFs. These GCSs play fairly important roles in the reduction of PEP upper bounds of transmitted signals in orthogonal frequency-division multiplexing (OFDM) communication systems. In general, GCSs deduced from GBFs have good algebraic constructions, and larger family sizes than the ones from other methods, which is advantageous to improvement of code rates in an OFDM communication system. However, a drawback that resultant sequences' length must be a power of 2 is obvious as well.

In this chapter, GBFs will be introduced, and a overview with regard to constructions of GCSs from GBFs will be given.

2. PERMUTATIONS

In this section, definitions of mappings and permutations will be given. For the readers who are familiar with these concepts, please skip them.

Definition 1. *Let A and B be two sets. Define a rule f from A to B. For arbitrarily given an element a in A, there is sole an element $b \in B$ found by the rule f, which is denoted by $b = f(a)$. Then, this rule f is referred to as a mapping from A to B. Frequently, a mapping f is expressed by*

$$f : \begin{array}{l} A \to B \\ a \to b = f(a), \end{array} \tag{1}$$

and b is called an image of a under mapping f, and a is said to be an original image of b.

Example 2.1. Let $A = \{a, b, c\}$ and $B = \{x, y, z\}$. Define

$$f_1(a) = x, \ f_1(b) = x, \ \text{and}, \ f_1(c) = y$$
$$f_2(a) = x, \ f_2(b) = y, \ \text{and}, \ f_2(c) = z$$
$$f_3(a) = x, \ f_3(a) = z, \ \text{and}, \ f_3(c) = y.$$

Then, f_1 and f_2 are two mappings from A to B, but not for f_3.

Definition 2. *Let f be a mapping from A to B. For $\forall a, b \in A$, $a \neq b$ such that the images of a and b under the mapping f must be distinct, that is, $f(a) \neq f(b)$, then, the mapping f is referred to as an injective mapping.*

Apparently, in Example 2.1, the mapping f_2 is an injective mapping, but not for the mapping f_1.

Definition 3. *Let f be a mapping from A to B. For $\forall y \in B$, there exists at least an element $a \in A$ such that $f(a) = y$, then, this mapping f is called an surjective mapping.*

Example 2.2. Let $A = \{a, b, c, d\}$ and $B = \{x, y, z\}$. Define

$$f_4(a) = x, \quad f_4(b) = y, \quad f_4(c) = z, \quad \text{and,} \quad f_4(d) = z$$
$$f_5(a) = x, \quad f_5(b) = x, \quad f_5(c) = y, \quad \text{and,} \quad f_5(d) = y.$$

Then, f_4 is a surjective mapping from A to B, but not for f_5.

Definition 4. *Let f be a mapping from A to B. If the mapping f is injective and surjective synchronously, this mapping f is named by a bijective mapping.*

According to the definition of bijective mapping, the mapping f_2 in Example 2.1 is a bijective mapping.

Definition 5. *Let A be a set. A bijective mapping σ from A to A is referred to as a permutation of set A.*

Example 2.3. Let $A = \{1, 2, 3\}$. Define

$$\sigma_1(1) = 1, \quad \sigma_1(2) = 2, \quad \text{and,} \quad \sigma_1(3) = 3$$
$$\sigma_2(1) = 1, \quad \sigma_2(2) = 3, \quad \text{and,} \quad \sigma_2(3) = 2$$
$$\sigma_3(1) = 2, \quad \sigma_3(2) = 1, \quad \text{and,} \quad \sigma_3(3) = 3.$$

Then, σ_1, σ_2, and σ_3 are all permutations of set A.

Theorem 2.1. Let set A be finite with cardinality $|A|$. the number of permutations of set A is $|A|!$.

In Example 2.3, there totally exist $3! = 6$ permutations. The other three permutation are given as follows.

$$\sigma_4(1) = 2, \quad \sigma_4(2) = 3, \quad \text{and,} \quad \sigma_4(3) = 1$$
$$\sigma_5(1) = 3, \quad \sigma_5(2) = 1, \quad \text{and,} \quad \sigma_5(3) = 2$$
$$\sigma_6(1) = 3, \quad \sigma_6(2) = 2, \quad \text{and,} \quad \sigma_6(3) = 1.$$

3. GENERALIZED BOOLEAN FUNCTIONS

In this section, Boolean function and GBFs are defined, including deduced sequences from them.

3.1. Boolean Functions and Deduced Sequences

Definition 6. *Let m be a positive integer ($m \geq 1$). A m-dimensional Boolean function $f(x_1, x_2, \cdots, x_m)$ is in fact a mapping from Z_2^m to Z_2, where $x_i \in Z_2$ ($1 \leq i \leq m$).*

For easy expression, denote $\underline{x} = (x_1, x_2, \cdots, x_m)$. Thus, Boolean function $f(x_1, x_2, \cdots, x_m)$ is simply written by $f(\underline{x})$.

For example, the following functions are Boolean functions.

$$\begin{aligned} f_1(\underline{x}) &= 1 \\ f_2(\underline{x}) &= x_2 \\ f_3(\underline{x}) &= x_1 x_2 + x_3 x_4 \cdots x_m \ (m \geq 3). \end{aligned} \tag{2}$$

It should be noted that the following 2^m monomials are crucial, since any Boolean function can be uniquely expressed by a linear combination of these monomials, in which the coefficient of each monomial belongs to Z_2.

$$\begin{aligned} &1, \\ &x_1, x_2, \cdots, x_m, \\ &x_1 x_2, x_1 x_3, \cdots, x_{m-1} x_m, \\ &x_1 x_2 x_3, x_1 x_2 x_4, \cdots, x_{m-2} x_{m-1} x_m, \\ &\vdots \\ &x_1 x_2 \cdots x_m. \end{aligned} \tag{3}$$

For a given m-dimensional vector $(x_1, x_2, \cdots, x_m)$, Boolean function $f(x_1, x_2, \cdots, x_m)$ has a value in Z_2. When m-dimensional vector $(x_1, x_2, \cdots, x_m)$ ranges from $\overbrace{(0, \cdots, 0)}^{m}$ to $\overbrace{(1, \cdots, 1)}^{m}$, 2^m values of Boolean function $f(\underline{x})$ can be deduced. For instance, consider $m = 2$ and $f(\underline{x}) = x_1 + x_2$. When two dimensional vector (x_1, x_2) ranges from $(0, 0)$ to $(1, 1)$, the resultant four values of Boolean function $f(\underline{x})$ are given below.

$$f(0, 0) = 0, \, f(0, 1) = 1, \, f(1, 0) = 1, \, f(1, 1) = 0.$$

For convenience, these values are arranged in lexicographic order so as to form a binary sequence $\underline{f}$ of length $N = 4$ as follows.

$$\underline{f} = (f(0, 0), f(0, 1), f(1, 0), f(1, 1)) = (0, 1, 1, 0) \text{ or } (0110). \tag{4}$$

Note that for an arbitrary integer i ($0 \leq i \leq 2^m - 1$), i can be always and uniquely expressed by

$$i = \sum_{k=1}^{m} i_k 2^{m-k} \ (i_k \in Z_2, 1 \leq k \leq m), \tag{5}$$

where the m-dimensional vector $(i_1, i_2, \cdots, i_m)$ is referred to as the binary representation of the integer i. Thus, there exists a bijective mapping between an integer i and its binary representation $(i_1, i_2, \cdots, i_m)$, which frequently results in mixed use of an integer and its binary representation regardless of difference of both concepts. For example, $f(0,0)$ can be written as $f(0)$ in this sense. Thus, the aforementioned binary sequence f in (4) can be equivalently written as $f = (f(0), f(1), f(2), f(3)) = (0, 1, 1, 0)$. In general, for an arbitrary positive integer m, a Boolean function $f(x)$ can deduce a binary sequence

$$f = (f(0), f(1), f(2), \cdots, f(2^m - 1))$$

of length $N = 2^m$.

Example 3.1. Take $m = 3$, and employ Boolean functions f_1-f_3 in (2). Then, the binary sequences, deduced from these three Boolean functions, are given below.

$$\begin{aligned}
f_1 &= (f_1(0), f_1(1), f_1(2), f_1(3), f_1(4), f_1(5), f_1(6), f_1(7)) = (11111111) \\
f_2 &= (f_2(0), f_2(1), f_2(2), f_2(3), f_2(4), f_2(5), f_2(6), f_2(7)) = (00110011) \\
f_3 &= (f_3(0), f_3(1), f_3(2), f_3(3), f_3(4), f_3(5), f_3(6), f_3(7)) = (01010110).
\end{aligned}$$

3.2. Generalized Boolean Functions and Relevant Results

Definition 7. *Let m and H be two positive integers ($m \geq 1, H \geq 2$). An m-dimensional GBF $f(x_1, x_2, \cdots, x_m)$ is in fact a mapping from Z_2^m to Z_H, where $x_i \in Z_2$ ($1 \leq i \leq m$).*

Similar to Boolean functions, any GBF can be uniquely expressed by a linear combination of those monomials in (3). However, the coefficient of each monomial belongs to Z_H rather than Z_2. Additionally, every m-dimensional GBF can deduce a H-PSK sequence of length $N = 2^m$.

Example 3.2. Take $H = 4$ and $m = 3$. Define three GBFs g_1-g_3 from Z_2^3 to Z_4 as follows.

$$g_1(\underline{x}) = 2x_2$$
$$g_2(\underline{x}) = 3x_1x_2$$
$$g_3(\underline{x}) = x_1 + 3x_1x_2 + 2x_3.$$

Then, these three GBFs can deduce three quaternary or quadrature phase-shift keying (QPSK) sequences of length $N = 8$ below.

$$\underline{g}_1 = (g_1(0), g_1(1), g_1(2), g_1(3), g_1(4), g_1(5), g_1(6), g_1(7)) = (00220022)$$
$$\underline{g}_2 = (g_2(0), g_2(1), g_2(2), g_2(3), g_2(4), g_2(5), g_2(6), g_2(7)) = (00000033)$$
$$\underline{g}_3 = (g_3(0), g_3(1), g_3(2), g_3(3), g_3(4), g_3(5), g_3(6), g_3(7)) = (02021302).$$

GBFs are divided into standard and non-standard ones. In coming discussions, the sequences deduced from standard and non-standard GBFs have different performance at all.

3.2.1. Standard Generalized Boolean Functions

Definition 8. *Let m (≥ 1) and H (≥ 2) be two positive integers, and σ be a permutation of the symbol set $\{1, 2, \cdots, m\}$. Again set $c, c_k \in Z_H$ $(1 \leq k \leq m)$. Define an m-dimensional GBF:*

$$f(\underline{x}) = \frac{H}{2} \sum_{k=1}^{m-1} x_{\sigma(k)} x_{\sigma(k+1)} + \sum_{k=1}^{m} c_k x_k + c \qquad (6)$$

from Z_2^m to Z_H. Then, $f(\underline{x})$ is referred to as a standard GBF.

This definition is naturally generalized to the one in [4], and so is Theorem 3.1.

Theorem 3.1. The standard GBFs $f(\underline{x})$'s in (6) totally have $H^{m+1}m!/2$ $(m \geq 2)$.

Proof. According to Theorem 2.1, the permutations σ's in (6) totally have $m!$. On the other hand, for a known permutation σ of the set $\{1, 2, \cdots, m\}$, define a new permutation σ' such that $\sigma'(k) = \sigma(m + 1 - k)$ $(1 \leq k \leq m)$. Apparently, the mathematical expressions $\sum_{k=1}^{m-1} x_{\sigma(k)} x_{\sigma(k+1)}$ and $\sum_{k=1}^{m-1} x_{\sigma'(k)} x_{\sigma'(k+1)}$

are identical. Consequently, there are $m!/2$ different mathematical expressions $\sum\limits_{k=1}^{m-1} x_{\sigma(k)}x_{\sigma(k+1)}$'s when σ ranges over all permutations of the set $\{1, 2, \cdots, m\}$ taking on each permutation exactly once. Note that for a given permutation σ, different coefficients c and c_k's ($1 \leq k \leq m$) result in different functions $f(\underline{x})$'s, which totally have H^{m+1}. Summarizing the above, this theorem follows immediately. $\square$

Theorem 3.2. ([8], [9]) Let integers m, $H \geq 2$, $N = 2^m$, and other mathematical symbols be the same as ones in Definition 8. Consider two non-negative integers τ ($0 < \tau \leq N - 1$) and i ($0 \leq i \leq N - 1$). Set $l = i + \tau$ and $\xi = e^{2\pi j/H}$, where j stands for an imaginary unit, that is, $j^2 = -1$. If $i_{\sigma(1)} = l_{\sigma(1)}$, for $1 \leq k \leq m$ and given τ there must exist two integers i' ($0 \leq i' \leq N - 1$), and v ($1 \leq v \leq m$) satisfying

$$
\begin{cases}
v = \inf\{k | i_{\sigma(k)} \neq l_{\sigma(k)}, 1 \leq k \leq m\} & \text{(7a)} \\
i_{\sigma(k)} + l'_{\sigma(k)} = 1 \quad (k < v) & \text{(7b)} \\
l_{\sigma(k)} + i'_{\sigma(k)} = 1 \quad (k < v) & \text{(7c)} \\
i_{\sigma(k)} = i'_{\sigma(k)} \quad (k \geq v) & \text{(7d)} \\
l_{\sigma(k)} = l'_{\sigma(k)} \quad (k \geq v) & \text{(7e)} \\
\xi^{f(i') - f(l')} = -\xi^{f(i) - f(l)}, & \text{(7f)}
\end{cases}
$$

and there exists a 1-1 corresponding between the integer pairs (i, l) and (i', l'), where $l' = i' + \tau$, and the binary representations and their images of four integers i, i', l, and l' have the following relationship.

$$
\begin{aligned}
i &: (i_1, i_2, \cdots, i_m) \xrightarrow{\sigma} (i_{\sigma(1)}, i_{\sigma(2)}, \cdots, i_{\sigma(m)}) \\
l &: (l_1, l_2, \cdots, l_m) \xrightarrow{\sigma} (l_{\sigma(1)}, l_{\sigma(2)}, \cdots, l_{\sigma(m)}) \\
i' &: (i'_1, i'_2, \cdots, i'_m) \xrightarrow{\sigma} (i'_{\sigma(1)}, i'_{\sigma(2)}, \cdots, i'_{\sigma(m)}) \\
  &\qquad = (1 - i_{\sigma(1)}, \cdots, 1 - i_{\sigma(v-1)}, i_{\sigma(v)}, \cdots, i_{\sigma(m)}) \\
l' &: (l'_1, l'_2, \cdots, l'_m) \xrightarrow{\sigma} (l'_{\sigma(1)}, l'_{\sigma(2)}, \cdots, l'_{\sigma(m)}) \\
  &\qquad = (1 - l_{\sigma(1)}, \cdots, 1 - l_{\sigma(v-1)}, l_{\sigma(v)}, \cdots, l_{\sigma(m)}).
\end{aligned} \tag{8}
$$

Proof. Due to $i \neq l$ and $i_{\sigma(1)} = l_{\sigma(1)}$, there must exist a smallest integer v such that $i_{\sigma(v)} \neq l_{\sigma(v)}$, in other words,

$$
v = \inf\{k | i_{\sigma(k)} \neq l_{\sigma(k)}, 1 \leq k \leq m\}.
$$

Define two integers i' and l', the images of whose binary representations under permutation σ satisfy

$$\begin{cases} i'_{\sigma(k)} = 1 - i_{\sigma(k)} & k \leq v - 1 \\ i'_{\sigma(k)} = i_{\sigma(k)} & k \geq v \end{cases}$$

and

$$\begin{cases} l'_{\sigma(k)} = 1 - l_{\sigma(k)} & k \leq v - 1 \\ l'_{\sigma(k)} = l_{\sigma(k)} & k \geq v, \end{cases}$$

respectively. More clearly, we have

$$\begin{aligned} i' : (i'_1, i'_2, \cdots, i'_m) &\xrightarrow{\sigma} (i'_{\sigma(1)}, i'_{\sigma(2)}, \cdots, i'_{\sigma(m)}) \\ &= (1 - i_{\sigma(1)}, \cdots, 1 - i_{\sigma(v-1)}, i_{\sigma(v)}, \cdots, i_{\sigma(m)}) \\ l' : (l'_1, l'_2, \cdots, l'_m) &\xrightarrow{\sigma} (l'_{\sigma(1)}, l'_{\sigma(2)}, \cdots, l'_{\sigma(m)}) \\ &= (1 - l_{\sigma(1)}, \cdots, 1 - l_{\sigma(v-1)}, l_{\sigma(v)}, \cdots, l_{\sigma(m)}). \end{aligned}$$

Since $l = i + \tau$, $l' = i' + \tau$ from the definitions of i' and l' in (8). Additionally, it is apparent that there exists a bijective mapping between the integer pairs (i, l) and (i', l').

For $k \leq v - 1$, according to the definition of the integer v, we have

$$i_{\sigma(k)} = l_{\sigma(k)} \neq l'_{\sigma(k)} \text{ and } i'_{\sigma(k)} = l'_{\sigma(k)} \neq l_{\sigma(k)},$$

which results in

$$i_{\sigma(k)} + l'_{\sigma(k)} = 1 \ (k \leq v - 1)$$

and

$$l_{\sigma(k)} + i'_{\sigma(k)} = 1 \ (k \leq v - 1).$$

Besides, it is natural that (7d) and (7e) hold according to the definitions of i' and l'.

Notice that

$$f(i) - f(l) \overset{(6)}{=} \frac{H}{2} \sum_{k=1}^{m-1} \left[i_{\sigma(k)} i_{\sigma(k+1)} - l_{\sigma(k)} l_{\sigma(k+1)} \right] + \sum_{k=1}^{m} c_{\sigma(k)} \left[i_{\sigma(k)} - l_{\sigma(k)} \right]$$

$$\overset{\text{definition of } v}{=} \frac{H}{2} \sum_{k=v-1}^{m-1} \left[i_{\sigma(k)} i_{\sigma(k+1)} - l_{\sigma(k)} l_{\sigma(k+1)} \right] + \sum_{k=v}^{m} c_{\sigma(k)} \left[i_{\sigma(k)} - l_{\sigma(k)} \right]$$

With the same argument, we have

$$f(i') - f(l') \overset{(6)}{=} \frac{H}{2} \sum_{k=1}^{m-1} \left[i'_{\sigma(k)} i'_{\sigma(k+1)} - l'_{\sigma(k)} l'_{\sigma(k+1)} \right] + \sum_{k=1}^{m} c_{\sigma(k)} \left[i'_{\sigma(k)} - l'_{\sigma(k)} \right]$$

$$\overset{\text{definition of } v}{=} \frac{H}{2} \sum_{k=v-1}^{m-1} \left[i'_{\sigma(k)} i'_{\sigma(k+1)} - l'_{\sigma(k)} l'_{\sigma(k+1)} \right] + \sum_{k=v}^{m} c_{\sigma(k)} \left[i'_{\sigma(k)} - l'_{\sigma(k)} \right]$$

By employing (7d) and (7e), we have

$$
\begin{aligned}
&(f(i) - f(l)) - (f(i') - f(l')) \\
&= \frac{H}{2} \left[i_{\sigma(v-1)} i_{\sigma(v)} - i'_{\sigma(v-1)} i'_{\sigma(v)} - l_{\sigma(v-1)} l_{\sigma(v)} + l'_{\sigma(v-1)} l'_{\sigma(v)} \right] \\
&= \frac{H}{2} \left[i_{\sigma(v-1)} (i_{\sigma(v)} - l_{\sigma(v)}) - i'_{\sigma(v-1)} (i'_{\sigma(v)} - l'_{\sigma(v)}) \right] \\
&\overset{i_{\sigma(v)}=i'_{\sigma(v)} \text{ and } l_{\sigma(v)}=l'_{\sigma(v)}}{=} \frac{H}{2} \left[i_{\sigma(v-1)} - i'_{\sigma(v-1)} \right] \left[i_{\sigma(v)} - l_{\sigma(v)} \right] \\
&\overset{i'_{\sigma(v-1)}=1-i_{\sigma(v-1)}}{=} \frac{H}{2} \left[-1 + 2 i_{\sigma(v-1)} \right] \left[i_{\sigma(v)} - l_{\sigma(v)} \right].
\end{aligned}
$$

Due to that $i_{\sigma(v)} \neq l_{\sigma(v)}$ and $\xi^{H i_{\sigma(v-1)}(i_{\sigma(v)} - l_{\sigma(v)})} = 1$, we have

$$\xi^{(f(i)-f(l))-(f(i')-f(l'))} = \xi^{-j\pi(i_{\sigma(v)} - l_{\sigma(v)})} = -1,$$

which results in (7f) immediately. $\square$

3.2.2. Non-Standard Generalized Boolean Functions

The GBFs that cannot be expressed by (6) are called the non-standard ones. For this concept in itself, constructions of non-standard GBFs are not difficult. Whereas, when associated with applications, such as, constructions of GCSs over Z_4, such constructions are a great challenge. For application of quaternary GCSs, typical constructions to produce non-standard GBFs over Z_4 are proposed by [5] and [6], whose importance will be shown in coming sections.

Theorem 3.3. ([6], Th.10) Let the integer $m \geq 1$, and let the integers t and ℓ satisfy $0 \leq t \leq m$ and $2 \leq \ell \leq m$, respectively. Again let σ be a permutation from the set $\{1, 2, \cdots, m\}$ to the set $\{1, 2, \cdots, m + 3\} \backslash \{m - t + 1, m - t + 2, m - t + 3\}$. Then, for $\forall \, e_0, \cdots, e_m \in Z_4$ and for $\forall \, u_0, u_1, u_2, u_3 \in Z_2$, $(m + 3)$-dimensional non-standard GBF $f(x_1, x_2, \cdots, x_{m+3})$ over Z_4 can be constructed by any one of the four cases below.

Case 1.

$$f(x_1, x_2, \cdots, x_{m+3}) =$$
$$f_1(x_1, x_2, \cdots, x_{m+3}) + 2x_{m-t+2}x_{m-t+3} + 2u_0x_{m-t+1} \qquad (9)$$
$$+2u_3x_{m-t+2} + (2u_0 + 2u_2 + u_3)x_{m-t+3}.$$

Case 2.

$$f(x_1, x_2, \cdots, x_{m+3}) =$$
$$f_1(x_1, x_2, \cdots, x_{m+3}) + 2x_{m-t+1}x_{m-t+3} + 2x_{m-t+2}x_{\sigma(1)}$$
$$+2x_{m-t+3}x_{\sigma(1)} + (2u_1 + 2u_2 + 2u_3 + 1)x_{m-t+2}+ \qquad (10)$$
$$(2u_1 + u_3 + 1)x_{m-t+3}.$$

Case 3.

$$f(x_1, x_2, \cdots, x_{m+3}) =$$
$$f_2(x_1, x_2, \cdots, x_{m+3}) + 2x_{m-t+2}x_{m-t+3} + 2u_0x_{m-t+1} \qquad (11)$$
$$+2u_3x_{m-t+2} + (2u_0 + 2u_2 + u_3)x_{m-t+3}.$$

Case 4.

$$f(x_1, x_2, \cdots, x_{m+3}) =$$
$$f_2(x_1, x_2, \cdots, x_{m+3}) + 2x_{m-t+1}x_{m-t+3} + 2x_{m-t+2}x_{\sigma(\ell-1)}$$
$$+2x_{m-t+3}x_{\sigma(\ell-1)} + 2x_{m-t+1}x_{\sigma(\ell)} + 2x_{m-t+3}x_{\sigma(\ell)}+ \qquad (12)$$
$$(2u_1 + 2u_2 + 2u_3 + 1)x_{m-t+2} + (2u_1 + u_3 + 1)x_{m-t+3}.$$

In Cases 1-4, the functions $f_1(\cdot)$ and $f_2(\cdot)$ are given by

$$f_1(x_1, x_2, \cdots, x_{m+3}) = 2x_{m-t+1}x_{m-t+3}x_{\sigma(1)}+$$
$$2x_{m-t+2}x_{m-t+3}x_{\sigma(1)} + 2x_{m-t+1}x_{m-t+2} + 2u_0x_{m-t+1}x_{\sigma(1)}$$
$$+(2u_1 + 2u_2 + 1)x_{m-t+2}x_{\sigma(1)} + (2u_0 + 2u_1 + 2u_2 \qquad (13)$$
$$+1)x_{m-t+3}x_{\sigma(1)} + 2\sum_{k=1}^{m-1} x_{\sigma(k)}x_{\sigma(k+1)} + \sum_{k=1}^{m} e_k x_{\sigma(k)} + e_0$$

and

$$f_2(x_1, x_2, \cdots, x_{m+3}) = 2x_{m-t+1}x_{m-t+3}x_{\sigma(\ell-1)}+$$
$$2x_{m-t+2}x_{m-t+3}x_{\sigma(\ell-1)} + 2x_{m-t+1}x_{m-t+3}x_{\sigma(\ell)}+$$
$$2x_{m-t+2}x_{m-t+3}x_{\sigma(\ell)} + 2x_{m-t+1}x_{\sigma(\ell-1)}x_{\sigma(\ell)}+$$
$$2x_{m-t+3}x_{\sigma(\ell-1)}x_{\sigma(\ell)} + 2x_{m-t+1}x_{m-t+2}+$$
$$2u_0x_{m-t+1}x_{\sigma(\ell-1)} + (2u_1 + 2u_2 + 1)x_{m-t+2}x_{\sigma(\ell-1)}+$$
$$(2u_0 + 2u_1 + 2u_2 + 1)x_{m-t+3}x_{\sigma(\ell-1)} + 2u_0x_{m-t+1}x_{\sigma(\ell)} \qquad (14)$$
$$+(2u_1 + 2u_2 + 3)x_{m-t+2}x_{\sigma(\ell)} + (2u_0 + 2u_1 + 2u_2 + 3)\cdot$$
$$x_{m-t+3}x_{\sigma(\ell)} + 2\sum_{k=1,k\neq\ell-1}^{m-1} x_{\sigma(k)}x_{\sigma(k+1)} + \sum_{k=1}^{m} e_k x_{\sigma(k)} + e_0$$

respectively.

Example 3.3. Consider $m = t = \ell = 2$ in Theorem 3.3, and $\sigma(1) = 5$ and $\sigma(2) = 4$. Again take on (9) and (13) with $(u_0, u_1, u_2, u_3) = (0, 0, 0, 0), e = 0$, and $e_k = 0$ $(0 \leq k \leq m)$. Thus, a non-standard GBF is obtained by

$$f(x_1, x_2, x_3, x_4, x_5) = 2x_1x_2 + 2x_2x_3 + x_2x_5 + x_3x_5 + 2x_4x_5 + 2x_1x_3x_5 + 2x_2x_3x_5.$$
$$(15)$$

Theorem 3.4. ([6], Corollary 11) Let $n = m + 3$ and $m \geq 1$. There are $(n-2)!(n-2)4^n$ non-standard GBFs over Z_4 in Theorem 3.3.

4. GOLAY COMPLEMENTARY SEQUENCES AND CONTROL OF PEAK ENVELOPE POWER

In this section, definitions of GCSs and PEP will be introduced, including relationship between them.

Sequences are closely associated with signal constellations. For different signal constellations, design methods and performances of sequences are fairly different. In this chapter, relevant constellations mainly include binary, quaternary, and QAM ones.

Let H (≥ 2) be a positive integer. A polyphase constellation means a signal set:

$$\{\xi^a | \xi = e^{j\frac{2\pi}{H}}, a \in Z_H\}. \tag{16}$$

Notice that there exists a bijective mapping between ξ^a and a. Henceforth, polyphase constellation is frequently simplified into mathematical expression $Z_H = \{0, 1, 2, \cdots, H - 1\}$. In particular, when $H = 2$ or 4, corresponding constellations are referred to as binary or quaternary ones, namely, Z_2 or Z_4. For example, binary sequence $\underline{A}_1 = (1, -1, 1, 1)$ and quaternary sequence $\underline{A}_2 = (-1, -j, 1, j)$ of length $N = 4$ are briefly written by $\underline{A}_1 = (0100)$ and $\underline{A}_2 = (2301)$, respectively.

The QAM constellation with order 4^q (positive integer $q \geq 2$) is the following symbol set: [7]

$$\Omega_{4^q\text{-QAM}} \overset{def}{=} \{a + bj | -2^q + 1 \leq a, b \leq 2^q - 1, a, b \text{ odd}\}.$$

Simply, when q = 2, 3, and 4, 16-QAM (see Figure 1, please), 64-QAM, and 256-QAM constellations appear, respectively.

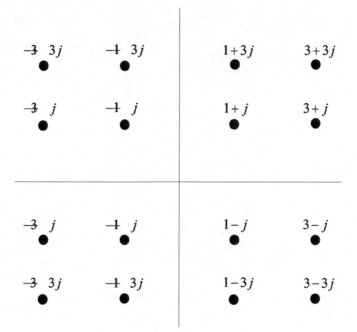

Figure 1. 16-QAM constellation.

4.1. Concepts of Golay Complementary Sequences

Definition 9. *Consider two complex sequences:* $\underline{A} = (A(0), A(1), \cdots, A(N-1))$ *and* $\underline{B} = (B(0), B(1), \cdots, B(N-1))$ *of length* N*. Define*

$$
C_{A,B}(\tau) = \begin{cases} \sum\limits_{i=0}^{N-1-\tau} A(i)\overline{B(i+\tau)} & 0 \leq \tau \leq N-1 \\ \sum\limits_{i=0}^{N-1+\tau} A(i-\tau)\overline{B(i)} & 1-N \leq \tau < 0 \\ 0 & |\tau| \geq N, \end{cases} \tag{17}
$$

where the symbol $\overline{x}$ *stands for the complex conjugation of* x*. If* $\underline{A} = \underline{B}$*,* $C_{A,A}(\tau)$ *is referred to as an aperiodic autocorrelation function (AACF), otherwise,* $C_{A,B}(\tau)$ *is said to be an aperiodic cross-correlation function (ACCF).*

It should be noted that when correlation function between sequences is calculated, components of sequences in Definition 9 don't allow to use simplified expressions.

An aperiodic correlation function has the following properties.

Theorem 4.1. For two complex sequences $\underline{A}$ and $\underline{B}$,

$$C_{A,B}(-\tau) = \overline{C_{B,A}(\tau)} \tag{18}$$

and

$$C_{A,A}(-\tau) = \overline{C_{A,A}(\tau)}. \tag{19}$$

Definition 10. *If two sequences $\underline{A}$ and $\underline{B}$ satisfy*

$$C_{A,A}(\tau) + C_{B,B}(\tau) = 0 \quad (\forall\, \tau \neq 0), \tag{20}$$

these two sequences are called Golay complementary sequences (GCSs) (also known as Golay complementary sequence pair (CSP) $(\underline{A}, \underline{B})$), and each of them is referred to as a Golay sequence.

Based on various signal constellations, GCSs are classified into binary, ternary, quaternary, polyphase, and QAM ones, etc.

Example 4.1. Take 16-QAM GCSs $\underline{A}$ and $\underline{B}$ of length $N = 16$ as follows.

$$
\begin{aligned}
\underline{A} = (&3-j, 1+3j, -1-3j, -3+j, 1-j, 1+j, 1+j, 1-j, -1-j, 1-j, \\
&-1+j, 1+j, 3+j, -1+3j, -1+3j, 3+j) \\
\underline{B} = (&3-j, 1+3j, -1-3j, -3+j, 1-j, 1+j, 1+j, 1-j, 1+j, -1+j, \\
&1-j, -1-j, -3-j, 1-3j, 1-3j, -3-j).
\end{aligned}
$$

The sum of AACFs of these two sequences is depicted in Figure 2, which apparently is impulse-like.

4.2. Peak Envelope Power

In communications, orthogonal frequency-division multiplexing (OFDM) communication systems are popular. However, high peak envelope power (PEP) of transmitted signals of OFDM systems brings a great challenge for these systems' implementations. As a result, control of PEP is one of all-important issues in OFDM design. To date, there have existed lots of methods to reduce PEP, therein, coded method is welcome [3]. The PEP of OFDM signals, encoded by GCSs, can be better controlled, which owes to complementary correlation property of GCSs (see Figure 2, please). By employing well-designed GCSs of

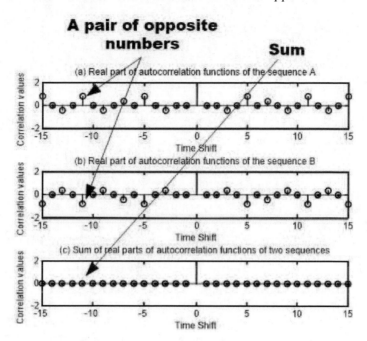

Figure 2. Sum of real parts of AACFs of 16-QAM GCSs $\underline{A}$ and $\underline{B}$.

length N over binary, quaternary, or polyphase constellations, upper bound of accompanying PEP does not exceed $2N$.

In an OFDM communication system of N sub-carriers, its i-th sub-carrier has the frequency f_i, where $f_i = f_0 + i\Delta f$ $(0 \leq i \leq N - 1)$, f_0 is the carrier frequency, and Δf is the spacing frequency between sub-carriers. A complex signal $s_A(t)$ $(0 \leq t \leq T_s)$, encoded by the sequence $\underline{A} = (A(0), A(1), \cdots, A(N - 1))$ of length N, is expressed by

$$s_A(t) = \sum_{k=0}^{N-1} A(k)e^{2\pi j f_k t}. \tag{21}$$

Thus, the transmitted OFDM signal in the OFDM system is exactly the real part of $s_A(t)$. Furthermore, the instantaneous envelope power $P_A(t)$ of the transmitted OFDM signal can be calculated by [8]

$$
\begin{aligned}
P_A(t) &= |s_A(t)|^2 \\
&= C_{A,A}(0) + \sum_{l \neq 0} C_{A,A}(l) e^{2\pi j l \Delta f t}.
\end{aligned} \tag{22}
$$

It has been proved that the instantaneous envelope power of an OFDM system is bounded by [4]

$$
P_A(t) \leq N^2, \tag{23}
$$

and the worst case can be attained when this system is uncoded, that is, $\underline{A} = \overbrace{(1, \cdots, 1)}^{N}$.

Let $\underline{A}$ and $\underline{B}$ form the GCSs. By making use of the definition of GCSs, apparently, the following equation holds.

$$
P_A(t) + P_B(t) = C_{A,A}(0) + C_{B,B}(0). \tag{24}
$$

Let C be a code whose codewords consist of sequences. Define PEPs of a sequence $\underline{A}$ and code C as follows.

$$
\begin{aligned}
\text{PEP}(\underline{A}) &= \sup_{t \in [0, T_s]} P_A(t) \\
\text{PEP}(C) &= \max\{\text{PEP}(\underline{A}) | \forall\, \underline{A} \in C\}.
\end{aligned} \tag{25}
$$

Furthermore, define a peak-to-mean envelope power ratio (PMEPR) of code C as follows.

$$
\text{PMEPR}(C) = \text{PEP}(C)/P_{av}(C), \tag{26}
$$

where $P_{av}(C)$ intends the mean envelope power of an OFDM signal averaged over all OFDM signals generated from a code C, that is,

$$
\begin{aligned}
P_{av}(C) &= \frac{1}{T_s} \sum_{\underline{A} \in C} p(\underline{A}) \int_{t=0}^{T_s} P_A(t) dt \\
&= \sum_{\underline{A} \in C} p(\underline{A}) C_{A,A}(0),
\end{aligned} \tag{27}
$$

in which $p(\underline{A})$ stands for the probability of transmitting the sequence $\underline{A}$, due to

$$
\frac{1}{T_s} \int_{t=0}^{T_s} P_A(t) dt = C_{A,A}(0). \tag{28}
$$

When an OFDM signal is encoded by well-designed binary, quaternary, or polyphase GCSs, the following theorem holds.

Theorem 4.2. ([4]) Let code C consist of binary, quaternary, or polyphase GCSs of length N. Then, the PEP of the code C satisfies

$$
\begin{aligned}
\text{PEP}(C) &\leq 2N \\
\text{PMEPR}(C) &\leq 2.
\end{aligned}
\tag{29}
$$

Theorem 4.2 is fit for binary, quaternary, and polyphase GCSs. However, for QAM GCSs, the conclusion on their PEP upper bound is given by

Theorem 4.3. ([9]) Let code C consist of 4^q-QAM general GCSs of length N from Cases I-III constructions in [9]. Then, the PEP of the code C satisfies

$$
\begin{aligned}
\text{PEP}(C) &\leq \frac{6N(2^q-1)}{2^q+1} \\
\text{PMEPR}(C) &\leq \frac{6(2^q-1)}{2^q+1}.
\end{aligned}
\tag{30}
$$

Here is Example 4.2 to highlight the importance of GCSs for reduction of PEP.

Example 4.2. Consider 4 sub-carriers system. Take the 16-QAM GCSs

$$
\begin{aligned}
\underline{A} &= \tfrac{1}{\sqrt{10}}(1+j, 1+j, 1+j, 1+j) \\
\underline{B} &= \tfrac{1}{\sqrt{10}}(1+j, -1+3j, 1-3j, -1-j)
\end{aligned}
$$

of length $N = 4$. Apparently, for the encoded OFDM signal, we have

$$
P_A(t) + P_B(t) = C_{A,A}(0) + C_{B,B}(0) = 3.2 = 0.8N.
$$

Figure 3 simulates the uncoded and encoded OFDM signals, respectively. Clearly, the uncoded OFDM signal has maximum magnitude 4. Whereas, the encoded OFDM signal is bounded to maximum magnitude 2. Hence, the improvement of PEP of the encoded OFDM system is obvious.

5. CONSTRUCTIONS OF 2^h-PSK GCSs

Binary GCSs can trace back to 1949, and were introduced by Golay to apply to infrared multislit spectrometry, optical time-domain reflectometry, and acoustic

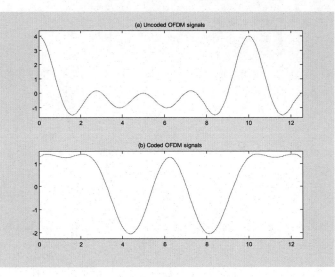

Figure 3. 4 sub-carriers uncoded and Coded OFDM signals.

surface-wave encoding, etc. It have been proven that lengths N's of all binary GCSs must be the forms $N = 2^{\alpha}10^{\beta}26^{\gamma}$'s (integers $\alpha, \beta, \gamma \geq 0$) [1]. However, non-binary GCSs have more other lengths. In 1999, Davis and Jedwab employed GBFs to construct 2^h-PSK GCSs (integer $h \geq 1$), which are referred to as 2^h-PSK GDJ GCSs. Apparently when $h = 1$ or 2, 2^h-PSK GDJ GCSs are in fact binary or quaternary GCSs. In this chapter, only construction methods of GCSs in connection with GBFs are investigated.

5.1.　2^h-PSK GDJ GCSs

Through this subsection, only standard GBFs are considered with $H = 2^h$ (integer $h \geq 1$) in Definition 8.

Theorem 5.1. ([4]) Take on a standard GBF $f(\underline{x})$ in (6). Construct two new GBFs below.

$$\begin{cases} a(\underline{x}) = f(\underline{x}) \\ b(\underline{x}) = f(\underline{x}) + 2^{h-1}x_{\sigma(1)} + c', \end{cases} \tag{31}$$

where $c' \in Z_{2^h}$. Write sequences $\underline{a}$ and $\underline{b}$ deduced from GBFs $a(\underline{x})$ and $b(\underline{x})$, respectively. Then, the sequences $\underline{a}$ and $\underline{b}$ of length $N = 2^m$ form 2^h-PSK GDJ

GCSs.

Proof. For clarity, three sequences $\underline{f}$, $\underline{a}$, and $\underline{b}$, deduced from three GBFs $f(\underline{x})$, $a(\underline{x})$, and $b(\underline{x})$, of length $N = 2^m$ are rewritten as follows.

$$\begin{aligned}
\underline{f} &= (f(0), f(1), \cdots, f(N-1)) \\
\underline{a} &= (a(0), a(1), \cdots, a(N-1)) \\
\underline{b} &= (b(0), b(1), \cdots, b(N-1)),
\end{aligned}$$

where $f(i) = f(i_1, i_2, \cdots, i_m)$, $a(i) = a(i_1, i_2, \cdots, i_m)$, $b(i) = b(i_1, i_2, \cdots, i_m)$, and $i = \sum_{k=1}^{m} i_k 2^{m-k}$.

According to Theorem 4.1, it is sufficient to investigate case $\tau > 0$. In order to derive this theorem, consider two cases: $m = 1$ and $m \geq 2$, respectively.

Case 1. $m = 1$.

In this Case, the GBFs $a(\underline{x})$ and $b(\underline{x})$ degenerate to

$$\begin{cases} a(\underline{x}) = c_1 x_1 + c \\ b(\underline{x}) = c_1 x_1 + 2^{h-1} x_1 + c', \end{cases} \tag{32}$$

due to the fact that only there exists a unit permutation $\sigma(1) = 1$ of the set $\{1\}$. Furthermore, when binary variable x_1 ranges from 0 to 1, the deduced sequences $\underline{a}$ and $\underline{b}$ of length $N = 2^m = 2$ are given as follows.

$$\begin{cases} \underline{a} = (c, c_1 + c) \\ \underline{b} = (c', c_1 + c' + 2^{h-1}), \end{cases} \tag{33}$$

According to (17), apparently, out-of-phase AACFs of $\underline{a}$ and $\underline{b}$ can be calculated by hand below.

$$C_{a,a}(1) = \xi^c \cdot \overline{\xi^{c_1+c}} = \overline{\xi^{c_1}} \tag{34}$$

and

$$C_{b,b}(1) = \xi^{c'} \cdot \overline{\xi^{(c_1+c'+2^{h-1})}} \xi^{2^{h-1}} \overset{=-1}{=} -\overline{\xi^{c_1}}, \tag{35}$$

which results in

$$C_{a,a}(1) + C_{b,b}(1) = 0, \tag{36}$$

in other words, the sequences $\underline{a}$ and $\underline{b}$ are 2^h-PSK GCSs.

Case 1. $m \geq 2$.

Let $\tau > 0$ and $l = i + \tau$ ($0 \leq i \leq N - 1 - \tau$). $(l_1, l_2, \cdots, l_m)$ is the binary representation of integer l. According to Definition 9, we calculate

$$
\begin{aligned}
& C_{a,a}(\tau) + C_{b,b}(\tau) \\
= & \sum_{i=0}^{N-1-\tau} \left[\xi^{a(i)-a(l)} + \xi^{b(i)-b(l)} \right] \\
= & \sum_{i=0}^{N-1-\tau} \xi^{b(i)-b(l)} \left[\xi^{a(i)-a(l)-b(i))+b(l)} + 1 \right]
\end{aligned}
\tag{37}
$$

Notice that

$$
\begin{aligned}
a(i) &= f(i) \\
a(l) &= f(l) \\
b(i) &= f(i) + 2^{h-1} i_{\sigma(1)} + c' \\
b(l) &= f(l) + 2^{h-1} l_{\sigma(1)} + c'.
\end{aligned}
\tag{38}
$$

As a result, (37) is simplified to

$$
C_{a,a}(\tau) + C_{b,b}(\tau) = \sum_{i=0}^{N-1-\tau} \xi^{f(i)-f(l)} \xi^{2^{h-1}(i_{\sigma(1)}-l_{\pi(1)})} \left[\xi^{2^{h-1}(l_{\sigma(1)}-i_{\sigma(1)})} + 1 \right].
\tag{39}
$$

Consider two cases: $i_{\sigma(1)} = l_{\sigma(1)}$ and $i_{\sigma(1)} \neq l_{\sigma(1)}$.

Case A. $i_{\sigma(1)} \neq l_{\sigma(1)}$.

This case implies $(i_{\sigma(1)}, l_{\sigma(1)}) = (1, 0)$ or $(i_{\sigma(1)}, l_{\sigma(1)}) = (0, 1)$, which always results in $\xi^{2^{h-1}(l_{\sigma(1)}-i_{\sigma(1)})} = e^{j\pi(l_{\sigma(1)}-i_{\sigma(1)})} = -1$, in other words, $\xi^{2^{h-1}(l_{\sigma(1)}-i_{\sigma(1)})} + 1 = 0$. Further, we have

$$
C_{a,a}(\tau) + C_{b,b}(\tau) = 0.
\tag{40}
$$

Thus, the crux to prove this theorem lies at that the sum vanishes when $i_{\sigma(1)} = l_{\sigma(1)}$.

Case B. $i_{\sigma(1)} = l_{\sigma(1)}$.

In this case, (39) is simplified into

$$
C_{a,a}(\tau) + C_{b,b}(\tau) = 2 \sum_{i=0}^{N-1-\tau} \xi^{f(i)-f(l)}.
\tag{41}
$$

Let integers i' and l' be defined by Theorem 3.2. Again let

$$
\begin{aligned}
I_\tau &= \{i | 0 \le i \le N - 1 - \tau\} \\
I_{1,\tau} &= \{i | i \text{ and } i' \text{ are defined by Theorem 3.2}, i \in I_\tau\} \\
I_{2,\tau} &= \{i' | i \text{ and } i' \text{ are defined by Theorem 3.2}, i \in I_\tau\},
\end{aligned}
\tag{42}
$$

which apparently satisfies

$$
\begin{cases}
I_\tau = I_{1,\tau} \bigcup I_{2,\tau} \\
I_{1,\tau} \bigcap I_{2,\tau} = \phi \text{ (empty set)}.
\end{cases}
\tag{43}
$$

Hence, (41) is simplified into

$$
\begin{aligned}
&C_{a,a}(\tau) + C_{b,b}(\tau) \\
&= 2 \sum_{i \in I_{1,\tau} \text{ and } i' \in I_{2,\tau}} \left[\xi^{f(i)-f(l)} + \xi^{f(i')-f(l')} \right] \\
&\overset{(7f)}{=} 0.
\end{aligned}
\tag{44}
$$

Henceforth, this theorem follows immediately. □

Theorem 5.2. 2^h-PSK GDJ GCSs totally have $2^{h(m+1)} m!/2$ $(m \ge 2)$.

Proof. Substitute $H = 2^h$ into Theorem 3.1, so this theorem is true. □

Example 5.1. Consider $h = 2$ and $m = 4$. Take on the unit permutation σ, i.e., $\sigma(i) = i$ $(1 \le i \le 4)$, $c' = 0$, and the GBF $f(\underline{x})$ below.

$$
f(\underline{x}) = 2 \sum_{k=1}^{3} x_k x_{k+1} + 2x_1 + x_2 + 3x_3 + x_4 + 3.
$$

Then, the resultant quaternary GDJ GCSs $\underline{a}$ and $\underline{b}$ of length $N = 2^m = 16$ by Theorem 5.1 are given as follows.

$$
\begin{aligned}
\underline{a} &= (3021011012030110) \\
\underline{b} &= (3021011030212332),
\end{aligned}
$$

sum of whose AACFs is

$$
C_{a,a}(\tau) + C_{b,b}(\tau) \overset{0 \le \tau \le 15}{=} (32, 0, 0, 0, 0, 0, 0, 0, 0, 0, 0, 0, 0, 0, 0, 0).
$$

5.2. Quaternary GCSs from Non-Standard GBFs

In this subsection, only quaternary GCSs constructed by non-standard GBFs are considered.

Theorem 5.3. ([6], Th.10) With the same conditions and the mathematical symbols as Theorem 3.3, Construct two new GBFs below.

$$\begin{cases} a(\underline{x}) = f(\underline{x}) \\ b(\underline{x}) = f(\underline{x}) + 2x_{\sigma(m)} + e, \end{cases} \tag{45}$$

where $e \in Z_4$. Then, the sequences $\underline{a}$ and $\underline{b}$, deduced from GBFs $a(\underline{x})$ and $b(\underline{x})$, respectively, of length $N = 2^{m+3}$ are quaternary GCSs.

    Number of non-standard quaternary GCSs from Theorem 5.3 is the same as one of non-standard GBFs, which is given in Theorem 3.4. For more constructions of non-standard quaternary GCSs, please refer to [6].

6. 4^q-QAM GCSs

In this section, only 4^q-QAM GCSs in connection with GBFs are discussed. The resultant sequences in this section are classified into 4^q-QAM GCSs based on standard binary GDJ GCSs, 4^q-QAM GCSs on basis of quaternary GDJ GCSs, and 4^q-QAM GCSs based on non-standard GBFs.

6.1. Descriptions of QAM Constellations

To date, the methods to produce 4^q-QAM constellations have been divided into Q-type and B-type descriptions. Q-type descriptions depend on quaternary inputs, and B-type ones are based on binary inputs. Figure 4 shows that diagrams of Q-type and B-type descriptions to produce QAM constellations.

6.1.1. Q-Type Descriptions

Q-type descriptions include Q-type-1 and Q-type-2 cases. In fact, Q-type-1 is a special case of Q-type-2. The Q-type-2 consists of $2^q - 1$ independent quaternary input variables, and is given by [10]

$$\left\{ (1+j) \sum_{p=0}^{2^q-2} j^{v^{(p)}} | v^{(p)} \in Z_4, 0 \le p \le 2^q - 2 \right\}. \tag{46}$$

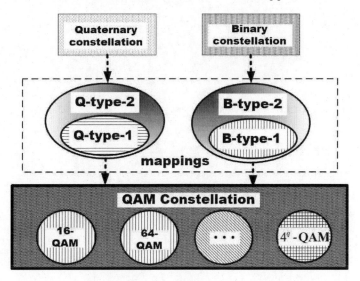

Figure 4. Q-type and B-type descriptions of QAM constellations.

Proof. In order to derive this construction, two cases need to be verified below.

(1) each symbol in Q-type-2 must belong to the set $\Omega_{4^q\text{-QAM}}$;

(2) each 4^q-QAM symbol can be produced by Q-type-2.

Case (1). For $\forall\ (v^{(0)}, v^{(1)}, \cdots, v^{(2^q-2)}) \in Z_4^{2^q-1}$ whose symbol distribution is given by

$$\begin{cases} n_0 = |\{p|v^{(p)} = 0, 0 \leq p \leq 2^q - 2\}| \\ n_1 = |\{p|v^{(p)} = 1, 0 \leq p \leq 2^q - 2\}| \\ n_2 = |\{p|v^{(p)} = 2, 0 \leq p \leq 2^q - 2\}| \\ n_3 = |\{p|v^{(p)} = 3, 0 \leq p \leq 2^q - 2\}|, \end{cases} \tag{47}$$

Q-type-2 description produces the symbol:

$$A = (1 + j) \sum_{p=0}^{2^q-2} j^{v^{(p)}} = (1 + j)\Big[(n_0 - n_2) + j(n_1 - n_3)\Big]$$
$$= (n_0 - n_2 - n_1 + n_3) + j(n_0 - n_2 + n_1 - n_3).$$

Notice that $n_0 + n_2 + n_1 + n_3 = 2^q - 1$ (odd number). Hence, in these four integers n_0, n_1, n_2, and n_3, only two cases: "one odd-three even" or "one even-three odd" appear. Clearly, both two cases result in the fact that both the integers "$n_0 - n_2 - n_1 + n_3$" and "$n_0 - n_2 + n_1 - n_3$" are odd. On the other hand, apparently, we have

$$-(2^q - 1) \leq n_0 - n_2 - n_1 + n_3, n_0 - n_2 + n_1 - n_3 \leq 2^q - 1.$$

Summarizing the above, we come to the conclusion: $A \in \Omega_{4^q\text{-QAM}}$.

Case (2). Consider $\forall\, a + jb \in \Omega_{4^q\text{-QAM}}$. We set

$$\sum_{k=0}^{2^q-2} j^{v^{(k)}} = \frac{a + jb}{1 + j} = \frac{a + b}{2} + j\frac{b - a}{2}. \tag{48}$$

Now, we need to determine the components of $(2^q - 1)$-dimensional vector $(v^{(0)}, v^{(1)}, \cdots, v^{(2^q-2)})$ over $Z_4^{2^q-1}$ to guarantee that (48) holds. A direct scheme is given below.

Step 1. Arbitrarily choose $\left|\frac{a+b}{2}\right|$ "0's" or "2's" in this vector, depending on $\frac{a+b}{2}$ positive or negative.

Step 2. Arbitrarily choose $\left|\frac{b-a}{2}\right|$ "1's" or "3's" in the other components except those components chosen in Step 1, depending on $\frac{b-a}{2}$ positive or negative.

Step 3. Arbitrarily and pairwise choose "0 and 2" or "1 and 3" in the unused components in Steps 1 and 2 so that the sum of powers of j from these unused components vanishes. To clarify this, we set $a = 2a_0 + 1$ and $b = 2b_0 + 1$ due to the fact that both a and b are odd, hence we have both $\frac{a+b}{2} = a_0 + b_0 + 1 = (b_0 - a_0) + (2a_0 + 1)$ and $\frac{b-a}{2} = b_0 - a_0$, which implies that $\left|\frac{a+b}{2}\right|$ and $\left|\frac{b-a}{2}\right|$ must be "one odd-one even". Therefore, the number $2^q - 1 - \left|\frac{a+b}{2}\right| - \left|\frac{b-a}{2}\right|$ of the unused components must be even, which means that didymous choice operation in Step 3 is feasible.

The above discussions suggest that for each in the set $\Omega_{4^q\text{-QAM}}$, there exists at least a $(2^q - 1)$-dimensional vector over $Z_4^{2^q-1}$ so that it is yielded by Q-type-2 description.

Thus, we verify that Q-type-2 description produce all the symbols in the set $\Omega_{4^q\text{-QAM}}$ when the input vector $(v^{(0)}, v^{(1)}, \cdots, v^{(2^q-2)}) \in Z_4^{2^q-1}$ ranges over the range from $(0, \cdots, 0)$ to $(3, \cdots, 3)$. $\square$

Note 1. Set

$$v^{(2^p-2+r)} = u^{(p)} \quad (0 \le p \le q-1, 1 \le r \le 2^p),$$

then Q-type-2 exactly degenerates to Q-type-1:

$$\left\{ (1+j) \sum_{p=0}^{q-1} 2^p j^{u^{(p)}} \middle| u^{(p)} \in Z_4, 0 \le p \le q-1 \right\}, \tag{49}$$

which consists of q independent quaternary input variables [7].

6.1.2. B-Type Descriptions

Similarly, B-type descriptions are divided into B-type-1 and B-type-2 cases, and B-type-1 is a special case of B-type-2. The B-type-2 consists of $2(2^q - 1)$ independent binary input variables, and is given by [11]

$$\left\{ \sum_{p=0}^{2^q-2} (-1)^{v_R^{(p)}} + j \sum_{p=0}^{2^q-2} (-1)^{v_I^{(p)}} \middle| v_R^{(p)}, v_I^{(p)} \in Z_2, 0 \le p \le 2^q - 2 \right\}.$$

Proof. This derivation of this construction is similar to one in Q-type-2. For completeness, we continue. Two cases must be investigated as follows.

Case 1. every symbol in B-type-2 must be in the set $\Omega_{4^q\text{-QAM}}$.

Case 2. every 4^q-QAM symbol can be yielded by B-type-2 description.

(1) For $\forall$ $(v_R^{(0)}, v_R^{(1)}, \cdots, v_R^{(2^q-2)}) \in Z_2^{2^q-1}$ and $\forall$ $(v_I^{(0)}, v_I^{(1)}, \cdots, v_I^{(2^q-2)}) \in Z_2^{2^q-1}$, their symbol distribution is given below.

$$\begin{cases} n_{R,0} = |\{p|v_R^{(p)} = 0, 0 \le p \le 2^q - 2\}| \\ n_{R,1} = |\{p|v_R^{(p)} = 1, 0 \le p \le 2^q - 2\}| \\ n_{I,0} = |\{p|v_I^{(p)} = 0, 0 \le p \le 2^q - 2\}| \\ n_{I,1} = |\{p|v_I^{(p)} = 1, 0 \le p \le 2^q - 2\}|. \end{cases}$$

Visibly, B-type-2 description gives the following symbol.

$$A = \sum_{p=0}^{2^q-2} (-1)^{v_R^{(p)}} + j \sum_{p=0}^{2^q-2} (-1)^{v_I^{(p)}}$$
$$= (n_{R,0} - n_{R,1}) + j(n_{I,0} - n_{I,1}).$$

Owing to both $n_{R,0} + n_{R,1} = 2^q - 1$ (odd number) and $n_{I,0} + n_{I,1} = 2^q - 1$ (odd number). Consequently, the integers $n_{R,0}$ and $n_{R,1}$ must be "one odd-one even", and so are the integers $n_{I,0}$ and $n_{I,1}$. Thus, both the differences "$n_{R,0} - n_{R,1}$" and "$n_{I,0} - n_{I,1}$" must be odd. Notice that we apparently have

$$-(2^q - 1) \le n_{R,0} - n_{R,1}, n_{I,0} - n_{I,1} \le 2^q - 1.$$

By summing up the above reasons, we come to the conclusion: $A \in \Omega_{4^q\text{-QAM}}$. So, we complete the verification of Case 1.

(2) For proving Case 2, we consider $\forall\, a + jb \in \Omega_{4^q\text{-QAM}}$. Provided we have

$$\sum_{p=0}^{2^q-2} (-1)^{v_R^{(p)}} + j \sum_{p=0}^{2^q-2} (-1)^{v_I^{(p)}} = a + jb. \tag{50}$$

The key point to be solved is that we need to find $(2^q - 1)$-dimensional vectors $(v_R^{(0)}, v_R^{(1)}, \cdots, v_R^{(2^q-2)})$ and $(v_I^{(0)}, v_I^{(1)}, \cdots, v_I^{(2^q-2)})$ over $Z_2^{2^q-1}$ to satisfy (50). Our scheme is given as follows.

Step 1. Arbitrarily choose $|a|$ "0's" or "1's" in the vector $(v_R^{(0)}, v_R^{(1)}, \cdots, v_R^{(2^q-2)})$, depending on a positive or negative.

Step 2. Arbitrarily choose $|b|$ "0's" or "1's" in the vector $(v_I^{(0)}, v_I^{(1)}, \cdots, v_I^{(2^q-2)})$, depending on b positive or negative.

Step 3. Pairwise and arbitrarily choose "0 and 1" in the unused components in Steps 1 and 2, respectively, so that the sum of 1 and -1 from these unused components vanishes. Notice that both a and b are odd. Thus, the number $2^q - 1 - |a|$ of the unused components is even, and so is the number $2^q - 1 - |b|$, which implies that didymous choice operation in Step 3 is feasible.

Summarizing the above, we come to the conclusion that for every symbol $a + jb$ in the set $\Omega_{4^q\text{-QAM}}$, there exist at least a pair $(2^q - 1)$-dimensional vectors over $Z_2^{2^q-1}$ so as to produce the given QAM symbol $a + jb$. In other words, we complete the verification of Case 2. $\square$

Note 2. Obviously, set

$$\begin{cases} v_R^{(2^p-2+r)} = u_R^{(p)} & (0 \le p \le q-1, 1 \le r \le 2^p) \\ v_I^{(2^p-2+r)} = u_I^{(p)} & (0 \le p \le q-1, 1 \le r \le 2^p), \end{cases} \tag{51}$$

then B-type-2 is exactly equivalent to B-type-1:

$$\left\{ \sum_{p=0}^{q-1} 2^p (-1)^{u_R^{(p)}} + j \sum_{p=0}^{q-1} 2^p (-1)^{u_I^{(p)}} \mid u_R^{(p)}, u_I^{(p)} \in Z_2, 0 \le p < q \right\}, \qquad (52)$$

which consists of $2q$ independent binary input variables [16].

6.2. 4^q-QAM GCSs from Quaternary GDJ GCSs

4^q-QAM GCSs are presented in this subsection, based on Q-type-2 description in (46). Since independent quaternary input variables in Q-Type-2 are more than ones in Q-Type-1, more 4^q-QAM GCSs arouse from employment of Q-Type-2.

Theorem 6.1. ([10]) Employ quaternary GBFs $f(\underline{x})$'s with $H = 4$ and $m \ge 2$ in Definition 8, including all mathematical symbols. Let

$$\begin{cases} a^{(0)}(\underline{x}) = a(\underline{x}) = f(\underline{x}) \\ b^{(0)}(\underline{x}) = a^{(0)}(\underline{x}) + \mu(\underline{x}) = a(\underline{x}) + \mu(\underline{x}) \\ a^{(1)}(\underline{x}) = a(\underline{x}) + s^{(1)}(\underline{x}) \\ b^{(1)}(\underline{x}) = a^{(1)}(\underline{x}) + \mu(\underline{x}) = a(\underline{x}) + s^{(1)}(\underline{x}) + \mu(\underline{x}) \\ \vdots \\ a^{(2^q-2)}(\underline{x}) = a(\underline{x}) + s^{(2^q-2)}(\underline{x}) \\ b^{(2^q-2)}(\underline{x}) = a^{(2^q-2)}(\underline{x}) + \mu(\underline{x}) = a(\underline{x}) + s^{(2^q-2)}(\underline{x}) + \mu(\underline{x}). \end{cases} \qquad (53)$$

By employing Q-type-2 description in (46), construct 4^q-QAM sequences $\underline{A} = (A(0), A(1), \cdots, A(N-1))$ and $\underline{B} = (B(0), B(1), \cdots, B(N-1))$ with length $N = 2^m$ by

$$\begin{cases} A(i) = \delta\gamma \sum_{p=0}^{2^q-2} j^{a^{(p)}(i)} \\ B(i) = \delta\gamma \sum_{p=0}^{2^q-2} j^{b^{(p)}(i)}, \end{cases} \qquad (54)$$

where $\delta = \dfrac{1}{\sqrt{(4^q-1)/3}}$ and $\gamma = e^{\pi j/4}$. Then, the proposed QAM sequences $\underline{A}$ and $\underline{B}$ are 4^q-QAM GCSs if the offsets $s^{(p)}(\underline{x})$ $(1 \le p \le 2^q - 2)$ and the pairing difference $\mu(\underline{x})$ satisfy one of the following cases, where $d_l^{(p)} \in Z_4$ $(1 \le p \le 2^q - 2, 0 \le l \le 2)$.

Case I.

$$\mu(\underline{x}) = 2x_{\sigma(m)}$$
$$s^{(p)}(\underline{x}) = d_0^{(p)} + d_1^{(p)} x_{\sigma(1)} \ (1 \le p \le 2^q - 2). \tag{55}$$

Case II.

$$\mu(\underline{x}) = 2x_{\sigma(1)}$$
$$s^{(p)}(\underline{x}) = d_0^{(p)} + d_1^{(p)} x_{\sigma(m)} \ (1 \le p \le 2^q - 2). \tag{56}$$

Case III.

$$\mu(\underline{x}) = 2x_{\sigma(1)} \text{ or } 2x_{\sigma(m)}$$
$$s^{(p)}(\underline{x}) = d_0^{(p)} + d_1^{(p)} x_{\sigma(w)} + d_2^{(p)} x_{\sigma(w+1)}$$
$$\text{with} \quad 2d_0^{(p)} + d_1^{(p)} + d_2^{(p)} \equiv 0 \ (\text{mod } 4) \tag{57}$$
$$(1 \le p \le 2^q - 2, 1 \le w \le m - 1).$$

Proof. For $\forall \ \tau > 0$, the AACF of the sequence $\underline{A}$ in (54) is counted by

$$\frac{1}{\delta^2} C_{A,A}(\tau) = \sum_{i=0}^{N-\tau-1} \left(\sum_{p=0}^{2^q-2} j^{a^{(p)}(i)} \right) \left(\sum_{r=0}^{2^q-2} j^{-a^{(r)}(l)} \right)$$
$$= \sum_{p=0}^{2^q-2} \sum_{r=0}^{2^q-2} C_{a^{(p)},a^{(r)}}(\tau)$$
$$= \sum_{p=0}^{2^q-2} C_{a^{(p)},a^{(p)}}(\tau) + \sum_{\substack{0 \le p,r \le 2^q-2 \\ p \ne r}} C_{a^{(p)},a^{(r)}}(\tau),$$

where $l = i + \tau$.

Similarly, we have

$$\frac{1}{\delta^2} C_{B,B}(\tau) = \sum_{p=0}^{2^q-2} C_{b^{(p)},b^{(p)}}(\tau) + \sum_{\substack{0 \le p,r \le 2^q-2 \\ p \ne r}} C_{b^{(p)},b^{(r)}}(\tau).$$

According to Theorem 5.1, the sequences $\underline{a}^{(p)}$ and $\underline{b}^{(p)}$ $(0 \le p \le 2^q - 2)$ form quaternary GCSs. Hence, in order to guarantee that the sequences $\underline{A}$ and $\underline{B}$ are GCSs, it is sufficient for us to have

$$\sum_{\substack{0 \le p,r \le 2^q-2 \\ p \ne r}} \left(C_{a^{(p)},a^{(r)}}(\tau) + C_{b^{(p)},b^{(r)}}(\tau) \right) = 0 \ (\forall \ \tau > 0). \tag{58}$$

Denote the left-hand side of (58) as $Lhs(\tau)$. Further, $Lhs(\tau)$ can be equivalently expressed by

$$
\begin{aligned}
Lhs(\tau) = \sum_{r=1}^{2^q-2} & \Big[C_{a^{(0)},a^{(r)}}(\tau) + C_{b^{(0)},b^{(r)}}(\tau) \\
& + C_{a^{(r)},a^{(0)}}(\tau) + C_{b^{(r)},b^{(0)}}(\tau) \Big] + \sum_{1\le p<r\le 2^q-2} \Big[C_{a^{(p)},a^{(r)}}(\tau) \\
& + C_{b^{(p)},b^{(r)}}(\tau) + C_{a^{(r)},a^{(p)}}(\tau) + C_{b^{(r)},b^{(p)}}(\tau) \Big].
\end{aligned}
\tag{59}
$$

Note that we have

$$
j^{a^{(p)}(i)-a^{(r)}(l)} = j^{a(i)+s^{(p)}(i)-a(l)-s^{(r)}(l)} = j^{a(i)-a(l)} j^{s^{(p)}(i)-s^{(r)}(l)}
$$

and

$$
j^{b^{(p)}(i)-b^{(r)}(l)} = j^{a(i)+s^{(p)}(i)+\mu(i)-a(l)-s^{(r)}(l)-\mu(l)} = j^{a(i)-a(l)} j^{s^{(p)}(i)-s^{(r)}(l)} j^{\mu(i)-\mu(l)},
$$

where $s^{(0)}(i) = s^{(0)}(l) = 0$.

According to (17), by substituting just above two equations into (59), $Lhs(\tau)$ can be equivalently simplified as

$$
Lhs(\tau) = \sum_{r=1}^{2^q-2} h_1(\tau,r) + \sum_{1\le p<r\le 2^q-2} h_2(\tau,p,r),
\tag{60}
$$

where

$$
h_1(\tau,r) = \sum_{i=0}^{N-\tau-1} j^{a(i)-a(l)} \Big[j^{s^{(r)}(i)} + j^{-s^{(r)}(l)} \Big] \Big[1 + j^{\mu(i)-\mu(l)} \Big]
\tag{61}
$$

and

$$
h_2(\tau,p,r) = \sum_{i=0}^{N-\tau-1} j^{a(i)-a(l)} \Big[j^{s^{(p)}(i)-s^{(r)}(l)} + j^{s^{(r)}(i)-s^{(p)}(l)} \Big] \Big[1 + j^{\mu(i)-\mu(l)} \Big].
\tag{62}
$$

Case III.

Due to $j^{\mu(i)-\mu(l)} = (-1)^{i_{\sigma(1)}-l_{\sigma(1)}}$, for $\forall\, \tau(>0), p, r$ both $h_1(\tau,r) = 0$ and $h_2(\tau,p,r) = 0$ hold whenever $i_{\sigma(1)} \neq l_{\sigma(1)}$. Therefore, the crux to verify (58) is that for $\forall\, \tau(>0), p, r$, both functions: $h_1(\tau,r)$ and $h_2(\tau,p,r)$ vanish simultaneously whenever $i_{\sigma(1)} = l_{\sigma(1)}$.

Similar to Theorem 5.1, for given $\tau > 0$ let

$$
\begin{aligned}
I_\tau &= \{i | i_{\sigma(1)} = l_{\sigma(1)}, 0 \le i \le N - \tau - 1\} \\
I_{1,\tau} &= \{i | (i, i') \text{ defined by Theorem 3.2}, 0 \le i, i' \le N - \tau - 1\} \\
I_{2,\tau} &= \{i' | (i, i') \text{ defined by Theorem 3.2}, 0 \le i, i' \le N - \tau - 1\},
\end{aligned}
\tag{63}
$$

which apparently results in

$$
\left\{
\begin{aligned}
I_\tau &= I_{1,\tau} \bigcup I_{2,\tau} \\
I_{1,\tau} &\bigcap I_{2,\tau} = \phi \text{ (empty set)},
\end{aligned}
\right.
\tag{64}
$$

Note that $1 + j^{\mu(i)-\mu(l)} = 2$ whenever $i_{\sigma(1)} = l_{\sigma(1)}$. Hence, the functions $h_1(\tau, r)$ and $h_2(\tau, p, r)$ are respectively reduced to

$$
\begin{aligned}
h_1(\tau, r) = 2 &\sum_{i \in I_{1,\tau}} j^{a(i)-a(l)} \left[j^{s^{(r)}(i)} + j^{-s^{(r)}(l)} \right] \\
+ 2 &\sum_{i' \in I_{2,\tau}} j^{a(i')-a(l')} \left[j^{s^{(r)}(i')} + j^{-s^{(r)}(l')} \right]
\end{aligned}
\tag{65}
$$

and

$$
\begin{aligned}
h_2(\tau, p, r) = 2 &\sum_{i \in I_{1,\tau}} j^{a(i)-a(l)} \left[j^{s^{(p)}(i)-s^{(r)}(l)} + j^{s^{(r)}(i)-s^{(p)}(l)} \right] \\
+ 2 &\sum_{i' \in I_{2,\tau}} j^{a(i')-a(l')} \left[j^{s^{(p)}(i')-s^{(r)}(l')} + j^{s^{(r)}(i')-s^{(p)}(l')} \right].
\end{aligned}
\tag{66}
$$

Notice that the offset $s^{(f)}(\underline{x})$ $(1 \le f \le 2^q - 2)$ in (57) satisfies one of the following two cases.

Case A. $w \ge v$.

According to (7d) and (7e) in Theorem 3.2, we have

$$
\left\{
\begin{aligned}
s^{(f)}(i) &= s^{(f)}(i') \\
s^{(f)}(l) &= s^{(f)}(l'),
\end{aligned}
\right.
\tag{67}
$$

which results in that

$$
j^{s^{(r)}(i)} + j^{-s^{(r)}(l)} = j^{s^{(r)}(i')} + j^{-s^{(r)}(l')}
\tag{68}
$$

and

$$
j^{s^{(p)}(i)-s^{(r)}(l)} + j^{s^{(r)}(i)-s^{(p)}(l)} = j^{s^{(p)}(i')-s^{(r)}(l')} + j^{s^{(r)}(i')-s^{(p)}(l')}
\tag{69}
$$

in (65) and (66).

By employing (7f), (68), and (69), it is apparent that both integers i and i' contribute zero to both the functions $h_1(\tau, r)$ and $h_2(\tau, p, r)$. After all the integer pairs (i, i')'s defined by Theorem 3.2 are used, the functions $h_1(\tau, r)$ and $h_2(\tau, p, r)$ vanish synchronously.

Case B. $w < v$.

According to (7b) and (7c) in Theorem 3.2, and (57), we have

$$
\begin{cases}
\begin{aligned}
s^{(f)}(i) + s^{(f)}(l') &= 2d_0^{(f)} + d_1^{(f)}[i_{\sigma(w)} + l'_{\sigma(w)}] + d_2^{(f)}[i_{\sigma(w+1)} + l'_{\sigma(w+1)}] \\
&= 2d_0^{(f)} + d_1^{(f)} + d_2^{(f)} \equiv 0 \;(\mathrm{mod}\; 4) \\
s^{(f)}(l) + s^{(f)}(i') &= 2d_0^{(f)} + d_1^{(f)}[l_{\sigma(w)} + i'_{\sigma(w)}] + d_2^{(f)}[l_{\sigma(w+1)} + i'_{\sigma(w+1)}] \\
&= 2d_0^{(f)} + d_1^{(f)} + d_2^{(f)} \equiv 0 \;(\mathrm{mod}\; 4).
\end{aligned}
\end{cases}
\tag{70}
$$

From (70), thus, we obtain

$$
\begin{cases}
j^{s^{(r)}(i)} = j^{-s^{(r)}(l')} \\
j^{-s^{(r)}(l)} = j^{s^{(r)}(i')} \\
j^{s^{(p)}(i)-s^{(r)}(l)} = j^{s^{(r)}(i')-s^{(p)}(l')} \\
j^{s^{(r)}(i)-s^{(p)}(l)} = j^{s^{(p)}(i')-s^{(r)}(l')},
\end{cases}
\tag{71}
$$

which results in that both (68) and (69) hold synchronously. With the same argument in Case A, then, the functions $h_1(\tau, r)$ and $h_2(\tau, p, r)$ must vanish synchronously.

Summarizing the above, we draw the conclusion that Case III follows immediately.

Cases I and II.

Employ the technique used in [8]. For Case II, both (68) and (69) must hold due to $w = m \geq v$. Go through the same discussions as Case A in Case III, the conclusion in Case II is true. For Case I, by employing the mapping $\sigma'(w) = m - 1 - \sigma(w)$, the conclusion in Case I follows immediately. $\square$

All solutions of (57) are listed in Table 1.

Example 6.1. Consider 64-QAM GCSs of length $N = 2^5 = 32$. Let $\sigma(i) = i$ $(1 \leq i \leq 5)$, $h = 2$, and $(c, c_1, c_2, c_3, c_4, c_5) = (1, 3, 0, 1, 2, 0)$. Consequently,

Table 1. All the solutions of congruence equation in (57)

$(d_0^{(p)}, d_1^{(p)}, d_2^{(p)})$ with $2d_0^{(p)} + d_1^{(p)} + d_2^{(p)} \equiv 0 \pmod 4$							
(0,0,0)	(0,1,3)	(0,2,2)	(0,3,1)	(1,0,2)	(1,1,1)	(1,2,0)	(1,3,3)
(2,0,0)	(2,1,3)	(2,2,2)	(2,3,1)	(3,0,2)	(3,1,1)	(3,2,0)	(3,3,3)

the standard GBFs are

$$f(\underline{x}) = 2\sum_{k=1}^{4} x_k x_{k+1} + 3x_1 + x_3 + 2x_4 + 1.$$

Take $\mu(\underline{x}) = 2x_1$, $w = 3$, $(d_0^{(1)}, d_1^{(1)}, d_2^{(1)}) = (2,3,1)$, and $(d_0^{(p)}, d_1^{(p)}, d_2^{(p)}) = (3,0,2)$ $(p = 2,3,4,5,6)$. As a result, the offsets are given by

$$\begin{cases} s^{(1)}(\underline{x}) = 2 + 3x_3 + x_4 \\ s^{(p)}(\underline{x}) = 3 + 2x_4 \ (2 \le p \le 6). \end{cases}$$

Thus, we have

$$\begin{cases} a^{(0)}(\underline{x}) = 2(x_1x_2 + x_2x_3 + x_3x_4 + x_4x_5) + 3x_1 + x_3 + \\ \qquad\qquad 2x_4 + 1 \\ b^{(0)}(\underline{x}) = a^{(0)}(\underline{x}) + 2x_1 \\ a^{(1)}(\underline{x}) = a^{(0)}(\underline{x}) + 2 + 3x_3 + x_4 \\ b^{(1)}(\underline{x}) = a^{(1)}(\underline{x}) + 2x_1 \\ a^{(p)}(\underline{x}) = a^{(0)}(\underline{x}) + 3 + 2x_4 \ (2 \le p \le 6) \\ b^{(p)}(\underline{x}) = a^{(p)}(\underline{x}) + 2x_1 \ (2 \le p \le 6). \end{cases}$$

Then, the resultant 64-QAM GCSs of length $N = 32$ are

$$\begin{aligned} \underline{A} = (&5 + 5j, 5 + 5j, 5 + 3j, -5 - 3j, -5 + 3j, -5 + 3j, \\ &5 - 5j, -5 + 5j, 5 + 5j, 5 + 5j, 5 + 3j, -5 - 3j, 5 - 3j, \\ &5 - 3j, -5 + 5j, 5 - 5j, 5 - 5j, 5 - 5j, 3 - 5j, -3 + 5j, \\ &3 + 5j, 3 + 5j, -5 - 5j, 5 + 5j, -5 + 5j, -5 + 5j, -3 + 5j, \\ &3 - 5j, 3 + 5j, 3 + 5j, -5 - 5j, 5 + 5j) \end{aligned}$$

and

$$\begin{aligned}
\underline{B} = (&5 + 5j, 5 + 5j, 5 + 3j, -5 - 3j, -5 + 3j, -5 + 3j, \\
&5 - 5j, -5 + 5j, 5 + 5j, 5 + 5j, 5 + 3j, -5 - 3j, 5 - 3j, \\
&5 - 3j, -5 + 5j, 5 - 5j, -5 + 5j, -5 + 5j, -3 + 5j, 3 - 5j, \\
&-3 - 5j, -3 - 5j, 5 + 5j, -5 - 5j, 5 - 5j, 5 - 5j, 3 - 5j, \\
&-3 + 5j, -3 - 5j, -3 - 5j, 5 + 5j, -5 - 5j),
\end{aligned}$$

the sum of whose AACFs is

$$\begin{aligned}
C_{A,A}(\tau) + C_{B,B}(\tau) = (&64, 0, 0, 0, 0, 0, 0, 0, 0, 0, 0, 0, 0, 0, \\
&0, 0, 0, 0, 0, 0, 0, 0, 0, 0, 0, 0, 0, 0, 0, 0, 0, 0),
\end{aligned}$$

It seems to be a little difficult to investigate family size of 4^q-QAM GCSs in Theorem 6.1. However, when the following two special cases are considered, such family size is accurately determined.

For the first special case, we investigate that Q-type-2 is replaced by Q-type-1. In fact, the resultant 4^q-QAM GCSs [9] are referred to as the ones from Cases I-III constructions [13]. We have the following theorem.

Theorem 6.2. ([9], Corollary 3) When Theorem 6.1 degenerates to Cases I-III constructions, that is, (53) and (54) in Theorem 6.1 are changed into

$$\begin{cases}
a^{(0)}(\underline{x}) = a(\underline{x}) = f(\underline{x}) \\
b^{(0)}(\underline{x}) = a^{(0)}(\underline{x}) + \mu(\underline{x}) = a(\underline{x}) + \mu(\underline{x}) \\
a^{(1)}(\underline{x}) = a(\underline{x}) + s^{(1)}(\underline{x}) \\
b^{(1)}(\underline{x}) = a^{(1)}(\underline{x}) + \mu(\underline{x}) = a(\underline{x}) + s^{(1)}(\underline{x}) + \mu(\underline{x}) \\
\vdots \\
a^{(q-1)}(\underline{x}) = a(\underline{x}) + s^{(q-1)}(\underline{x}) \\
b^{(q-1)}(\underline{x}) = a^{(q-1)}(\underline{x}) + \mu(\underline{x}) = a(\underline{x}) + s^{(q-1)}(\underline{x}) + \mu(\underline{x}),
\end{cases} \tag{72}$$

and

$$\begin{cases}
A(i) = \delta\gamma \sum_{p=0}^{q-1} 2^{q-1-p} j^{a^{(p)}(i)} \\
B(i) = \delta\gamma \sum_{p=0}^{q-1} 2^{q-1-p} j^{b^{(p)}(i)},
\end{cases} \tag{73}$$

respectively. Family size of the resultant 4^q-QAM GCSs is given by

$$\left[(m + 1)4^{2(q-1)} - (m + 1)4^{q-1} + 2^{q-1} \right] \times (m!/2)4^{m+1} \quad (m \geq 2, q \geq 2). \tag{74}$$

Proof. Let S_I, S_{II}, and S_{III} be the sets of compatible offset combinations $(s^{(1)}(\underline{x}), \cdots, s^{(q-1)}(\underline{x}))$ from Cases I-III in (55)-(57), respectively.

$$S_I = \{d_0^{(p)} + d_1^{(p)} x_{\sigma(1)} | 1 \le p \le q - 1\}, \tag{75}$$

$$S_{II} = \{d_0^{(p)} + d_1^{(p)} x_{\sigma(m)} | 1 \le p \le q - 1\}, \tag{76}$$

and

$$S_{III} = \{d_0^{(p)} + d_1^{(p)} x_{\sigma(w)} + d_2^{(p)} x_{\sigma(w+1)} | 2d_0^{(p)} + d_1^{(p)} + d_2^{(p)} \equiv 0 \,(\mathrm{mod}\ 4),$$
$$1 \le w \le m - 1, 1 \le p \le q-\}, \tag{77}$$

where $d_l^{(p)} \in Z_4$ ($1 \le p \le 2^q - 2$ and $0 \le l \le 2$).

Apparently, the number of 4^q-QAM GCSs equals

$$|S_I \bigcup S_{II} \bigcup S_{III}| \times (m!/2) 4^{m+1},$$

where $|S_I \bigcup S_{II} \bigcup S_{III}|$ means the cardinality of union of compatible offset combinations $(s^{(1)}(\underline{x}), s^{(2)}(\underline{x}), \cdots, s^{(q-1)}(\underline{x}))$ from Cases I-III.

Apparently, we have

$$|S_I| = |S_{II}| = 4^{2(q-1)}.$$

In order to calculate $|S_{III}|$, let

$$S_{III}^w = \{d_0^{(p)} + d_1^{(p)} x_{\sigma(w)} + d_2^{(p)} x_{\sigma(w+1)} | 2d_0^{(p)} + d_1^{(p)} + d_2^{(p)} \equiv 0 \,(\mathrm{mod}\ 4),$$
$$1 \le p \le q - 1\}$$

stand for S_{III} with given w ($1 \le w \le m - 1$). Then, we have

$$S_{III} = \bigcup_{w=1}^{m-1} S_{III}^w.$$

Notice that in each S_{III}^w, indenpendent coefficient vectors $(d_0^{(p)}, d_1^{(p)}, d_2^{(p)})$'s with the constraints $2d_0^{(p)} + d_1^{(p)} + d_2^{(p)} \equiv 0 \,(\mathrm{mod}\ 4)$ ($1 \le p \le q-1$) totally have $q - 1$, and the solutions of these constraints totally have 16 in table 1. Thus, the cardinality of each S_{III}^w is $16^{q-1} = 4^{2(q-1)}$. Additionally, due to

$$\left(S_{III}^1 \bigcup S_{III}^2 \bigcup \cdots \bigcup S_{III}^{w-1}\right) \bigcap S_{III}^w$$
$$= \{d_0^{(p)} + d_1^{(p)} x_{\sigma(w)} | (d_0^{(p)}, d_1^{(p)}) = (0,0), (2,0), (1,2), (3,2), 1 \le p \le q - 1\}.$$

Hence, we have

$$|S_{III}|$$
$$= |S_{III}^1 \bigcup S_{III}^2 \bigcup \cdots \bigcup S_{III}^{w-1}|$$
$$= |S_{III}^1| + |S_{III}^2| - |S_{III}^1 \cap S_{III}^2| + \cdots + |S_{III}^{w-1}| - |(S_{III}^1 \bigcup S_{IIi}^2 \bigcup \cdots \bigcup S_{III}^{w-1}) \cap S_{III}^w|$$
$$= (m-1)4^{2(q-1)} - (m-2)4^{q-1}.$$

Let $S_{I,II}$, $S_{I,III}$, $S_{II,III}$, and $S_{I,II,III}$ stand for be intersections of S_I, S_{II}, and S_{III} as indicated by the subscripts. Apparently,

$$S_{I,II} = \{d_0^{(p)} | d_0^{(p)} \in Z_4, 1 \le p \le q-1\}$$

$$S_{I,III} = \{d_0^{(p)} + d_1^{(p)} x_{\sigma(1)} | (d_0^{(p)}, d_1^{(p)}) = (0,0), (2,0), (1,2), (3,2), 1 \le p \le q-1\}$$

$$S_{II,III} = \{d_0^{(p)} + d_1^{(p)} x_{\sigma(m)} | (d_0^{(p)}, d_1^{(p)}) = (0,0), (2,0), (1,2), (3,2), 1 \le p \le q-1\}$$

$$S_{I,II,III} = \{d_0^{(p)} | d_0^{(p)} \in \{0,2\}, 1 \le p \le q-1\}$$

Consequently, we have

$$|S_{I,II}| = |S_{I,III}| = |S_{II,III}| = 4^{q-1} \text{ and } |S_{I,II,III}| = 2^{q-1}.$$

Summarizing the above, we have

$$|S_I \bigcup S_{II} \cdots \bigcup S_{III}|$$
$$= |S_I| + |S_{II}| + |S_{III}| - |S_{I,II}| - |S_{I,III}| - |S_{II,III}| + |S_{I,II,III}|$$
$$= 4^{2(q-1)} + 4^{2(q-1)} + (m-1)4^{2(q-1)} - (m-2)4^{q-1} - 3 \cdot 4^{q-1} + 2^{q-1}$$
$$= (m+1)4^{2(q-1)} - (m+1)4^{q-1} + 2^{q-1}.$$

This theorem is verified. $\square$

For the second special case, only 16-QAM GCSs in Theorem 6.1 are considered. Then, we have the following theorem.

Theorem 6.3. ([10]) The number of 16-QAM GCSs of length 2^m in Theorem 6.1 is

$$(58m + 105)(m!/2)4^{m+1} \ (m \ge 2). \tag{78}$$

Proof. In this case, the construction in (54) is simplified as

$$\begin{cases} A(i) = \delta\gamma\left(j^{a^{(0)}(i)} + j^{a^{(0)}(i)+s^{(1)}(i)} + j^{a^{(0)}(i)+s^{(2)}(i)}\right) \\ B(i) = \delta\gamma\left(j^{a^{(0)}(i)+\mu(i)} + j^{a^{(0)}(i)+s^{(1)}(i)+\mu(i)} + j^{a^{(0)}(i)+s^{(2)}(i)+\mu(i)}\right), \end{cases} \tag{79}$$

which implies that for given the GBF $f(\underline{x})$ and the pair difference $\mu(\underline{x})$, the 16-QAM Golay CSP $(\underline{A}, \underline{B})$ is determined, only depending on the offset pair $(s^{(1)}(\underline{x}), s^{(2)}(\underline{x}))$. As a result, it is sufficient to calculate the number of the offset pairs for the family size of the proposed 16-QAM GCSs. Notice that for two arbitrary offset pairs $(g(\underline{x}), k(\underline{x}))$ and $(u(\underline{x}), v(\underline{x}))$, when both $g(\underline{x}) = v(\underline{x})$ and $k(\underline{x}) = u(\underline{x})$ appear synchronously, the two 16-QAM Golay CSPs, deduced by these two offset pairs referred to above, are identical. Such two offset pairs are called *overlapping*. Hence, the crux to calculate the family size is to count the number of non-overlapping offset pairs. Consider two cases below.

(1). $s^{(1)}(\underline{x}) = s^{(2)}(\underline{x})$.

In this case, the construction in (54) degenerates to the one in [8], namely, (72) and (73). By using the results in Theorem 6.2, we have $12m + 14$ non-overlapping offset pairs.

(2). $s^{(1)}(\underline{x}) \neq s^{(2)}(\underline{x})$.

Employing the method in [8] so as to divide possible offsets into the following five sets with empty pairwise intersection. Combining the possible offset coefficients $(d_0^{(p)}, d_1^{(p)}, d_2^{(p)})$'s which are listed in Table 1, we have

$$
\begin{aligned}
S_1 &= \{d_0 | d_0 = 0, 1, 2, 3\}; \\
S_2 &= \{d_0 + d_1 x_{\sigma(1)} | d_0, d_1 \in Z_4, d_1 \neq 0\}; \\
S_3 &= \{d_0 + d_1 x_{\sigma(m)} | d_0, d_1 \in Z_4, d_1 \neq 0\}; \\
S_4 &= \{d_0 + d_1 x_{\sigma(w)} | (d_0, d_1) = (1, 2), (3, 2)\} \, (2 \leq w \leq m - 1); \\
S_5 &= \{d_0 + d_1 x_{\sigma(w)} + d_2 x_{\sigma(w+1)} | (d_0, d_1, d_2) = (0, 1, 3), \\
&\quad (0, 2, 2), (0, 3, 1), (1, 1, 1), (1, 3, 3), (2, 1, 3), (2, 2, 2), \\
&\quad (2, 3, 1), (3, 1, 1), (3, 3, 3)\} \, (1 \leq w \leq m - 1),
\end{aligned}
$$

where the sets S_2 and S_3 belong to Cases 1 and 2, respectively, and the sets S_1, S_4, and S_5 are in Case 3 in Theorem 6 synchronously.

In order to obtain non-overlapping offset pairs $(s^{(1)}(\underline{x}), s^{(2)}(\underline{x}))$'s, we use the following selection strategy. Consider selection of offset pairs in S_i ($1 \leq i \leq 5$). Step 1: arbitrarily choose an offset in S_i, written by Ξ_1, as $s^{(1)}(\underline{x})$, and arbitrarily take an offset in $S_i - \Xi_1$ for $s^{(2)}(\underline{x})$, which can produce $|S_i| - 1$ possible offset pairs. Step 2: in $S_i - \Xi_1$, arbitrarily choose an offset (written by Ξ_2), and arbitrarily take an offset in $S_i - \Xi_1 - \Xi_2$ for $s^{(2)}(\underline{x})$, which can produce $|S_i| - 2$ possible offset pairs. So continue, until we have $|S_i - \Xi_1 - \Xi_2 - \cdots| \leq 1$. Summarizing up the above, we obtain $(|S_i| - 1) + (|S_i| - 2) + \cdots + 1$ possible offset pairs with no overlaps to one another.

According to the above strategy, (a) in S_1, we have 3+2+1=6 possible offset pairs; (b) in S_2, we have $11 + 10 + \cdots + 1 = 66$ possible offsets pairs; (c) in S_3, this case is the same as Case (b); (d) in S_4, only one pair exists. However, the parameter w can vary from 2 to $m - 1$. Hence, we have $m - 2$ possible offset pairs; (e) in S_5, there are $9 + 8 + \cdots + 1 = 45$ possible offset pairs, in which each has the parameter w that can vary from 1 to $m - 1$. Thus, this case has $45(m - 1)$ pairs in total. Summarizing up (a)-(e), we totally have $46m + 91$ possible offset pairs.

Summing up Cases (1) and (2), we come to the conclusion that the total number of non-overlapping offset pairs is $(12m + 14) + (46m + 91) = 58m + 105$. According to Theorem 3.1, hence, this theorem follows immediately. □

For higher order QAM constellations, in general, family size of 4^q-QAM GCSs in Theorem 6.1 is open.

Theorem 6.4. Let the code C consist of 4^q-QAM GCSs from Theorem 6.1 with length $N = 2^m$. Then, the PEP of the code C is bounded by

$$\text{PEP}(C) \leq \frac{6N(2^q - 1)}{2^q + 1}. \tag{80}$$

Proof. For $\forall \underline{A} \in C$, let $(\underline{A}, \underline{B})$ is a 4^q-QAM Golay CSP from Theorem 6.1 with length $N = 2^m$. According to (24), the PEP of the sequence $\underline{A}$ satisfies

$$\text{PEP}(\underline{A}) \leq 2C_{A,A}(0) = 2 \sum_{i=0}^{N-1} |A(i)|^2$$

From (54), we have

$$|A(i)|^2 = \left| \delta \sum_{p=0}^{2^q-2} j^{a^{(p)}(i)} \right|^2 \leq \delta^2 \sum_{p=0}^{2^q-2} 1 = \frac{3(2^q - 1)}{2^q + 1}. \tag{81}$$

According to (25), this theorem follows immediately. □

6.3. 4^q-QAM GCSs on Basis of Binary GDJ GCSs

It is advantageous for 4^q-QAM GCSs from B-type descriptions to apply to such QAM systems driven only by binary signals, such as in [14]. [11] is based on B-type-2 description, and [15] is from B-type-1 one. In general, the number of resultant 4^q-QAM GCSs from B-type-2 is larger than one for B-type-1.

Theorem 6.5. ([15]) Let integers n and m satisfy $n, m \geq 2$. Again let

$$
\begin{cases}
a(\underline{x}) = \sum\limits_{k=1}^{m-1} x_{\sigma(k)} x_{\sigma(k+1)} + \sum\limits_{k=1}^{m} e_k x_k + e \\
a^{(r)}(\underline{x}) = a(\underline{x}) + s^{(r)}(\underline{x}) \\
b^{(r)}(\underline{x}) = a^{(r)}(\underline{x}) + \mu(\underline{x}),
\end{cases}
\tag{82}
$$

where $0 \leq r \leq 2n - 1$, $e, e_k \in Z_2$ $(1 \leq k \leq m)$.

For $1 \leq w \leq m$, the offsets $s^{(r)}(\underline{x})$'s and the pairing difference $\mu(\underline{x})$ are given by one of the following cases.

Case I.

$$
\begin{cases}
s^{(r)}(\underline{x}) = d_{r0} + d_{r1} x_{\sigma(w)} + d_{r2} x_{\sigma(w+1)} \\
\mu(\underline{x}) = x_{\sigma(1)} \text{ or } x_{\sigma(m)},
\end{cases}
\tag{83}
$$

where $d_{rl} \in Z_2$ $(0 \leq r \leq 2n - 1$ and $0 \leq l \leq 2)$, and

$$
\begin{cases}
d_{01} + d_{02} + d_{h1} + d_{h2} \equiv 0 \pmod{2} & (1 \leq h \leq n - 1) \\
d_{01} + d_{02} + d_{(n+h)1} + d_{(n+h)2} \equiv 1 \pmod{2} & (0 \leq h \leq n - 1),
\end{cases}
\tag{84}
$$

Case II.

$$
\begin{cases}
s^{(r)}(\underline{x}) = d_{r0} + d_{r1} x_{\sigma(1)} \ (d_{r0}, d_{r1} \in Z_2, 0 \leq r < 2n) \\
\mu(\underline{x}) = x_{\sigma(m)}.
\end{cases}
\tag{85}
$$

Case III.

$$
\begin{cases}
s^{(r)}(\underline{x}) = d_{r0} + d_{r1} x_{\sigma(m)} \ (d_{r0}, d_{r1} \in Z_2, 0 \leq r < 2n) \\
\mu(\underline{x}) = x_{\sigma(1)}.
\end{cases}
\tag{86}
$$

Construct the 2^{2n}-QAM sequences $\underline{A} = (A(0), A(1), \cdots, A(N-1))$ and $\underline{B} = (B(0), B(1), \cdots, B(N-1))$ with length $N = 2^m$ by

$$
\begin{cases}
A(i) = \lambda \left[\sum\limits_{f=0}^{n-1} (-1)^{c_f} 2^f (-1)^{a^{(f)}(i)} + j \sum\limits_{g=0}^{n-1} (-1)^{d_g} 2^g (-1)^{a^{(n+g)}(i)} \right] \\
B(i) = \lambda \left[\sum\limits_{f=0}^{n-1} (-1)^{c_f} 2^f (-1)^{b^{(f)}(i)} + j \sum\limits_{g=0}^{n-1} (-1)^{d_g} 2^g (-1)^{b^{(n+g)}(i)} \right],
\end{cases}
\tag{87}
$$

where $\lambda = \dfrac{1}{\sqrt{2(4^n-1)/3}}$, the coefficients c_i's and d_i's ($0 \le i \le n-1$) belong to Z_2, and are determined before the sequences $\underline{A}$ and $\underline{B}$ are produced.

Then, the proposed square-QAM sequences $\underline{A}$ and $\underline{B}$ form 2^{2n}-QAM GCSs.

Note 3. (87) uses a modified version of B-type-1. When all coefficients c_i's and d_i's ($0 \le i \le n-1$) take on zeros, this modified version exactly degenerates to B-type-1. These coefficients affect the PEPs of resultant QAM GCSs.

Proof. The idea to derive this theorem is similar to one in Theorem 6.1, therefore, a similar investigation needs to be done. For completeness, we continue to state it with a simplified derivation skeleton. From the definition of AACFs, for arbitrarily given $\tau > 0$ we have

$$
\begin{aligned}
\frac{1}{\lambda^2}C_{A,A}(\tau) =& \\
&\sum_{i=0}^{N-\tau-1}\Big[\sum_{f=0}^{n-1}(-1)^{c_f}2^f(-1)^{a^{(f)}(i)}+j\sum_{g=0}^{n-1}(-1)^{d_g}2^g(-1)^{a^{(n+g)}(i)}\Big] \\
&\cdot\Big[\sum_{h=0}^{n-1}(-1)^{c_h}2^h(-1)^{a^{(h)}(l)}-j\sum_{k=0}^{n-1}(-1)^{d_k}2^k(-1)^{a^{(n+k)}(l)}\Big] \\
=&\sum_{f=0}^{n-1}\sum_{h=0}^{n-1}(-1)^{c_f+c_h}2^{f+h}C_{a(f),a(h)}(\tau)-j\sum_{f=0}^{n-1}\sum_{k=0}^{n-1}(-1)^{c_f+d_k}2^{f+k}C_{a(f),a(n+k)}(\tau)+ \\
&j\sum_{g=0}^{n-1}\sum_{h=0}^{n-1}(-1)^{d_g+c_h}2^{g+h}C_{a(n+g),a(h)}(\tau)+\sum_{g=0}^{n-1}\sum_{k=0}^{n-1}(-1)^{d_g+d_k}2^{g+k}C_{a(n+g),a(n+k)}(\tau) \\
=&\Big[\sum_{\substack{0\le f\le n-1\\ h=f}}+\sum_{f=0}^{n-1}\sum_{\substack{0\le h\le n-1\\ h\ne f}}\Big](-1)^{c_f+c_h}2^{f+h}C_{a(f),a(h)}(\tau) \\
&-j\sum_{f=0}^{n-1}\sum_{h=0}^{n-1}(-1)^{c_f+d_h}2^{f+h}C_{a(f),a(n+h)}(\tau)+j\sum_{f=0}^{n-1}\sum_{h=0}^{n-1}(-1)^{d_f+c_h}2^{f+h}C_{a(n+f),a(h)}(\tau) \\
&+\Big[\sum_{\substack{0\le f\le n-1\\ h=f}}+\sum_{f=0}^{n-1}\sum_{\substack{0\le h\le n-1\\ h\ne f}}\Big](-1)^{d_f+d_h}2^{f+h}C_{a(n+f),a(n+h)}(\tau) \\
=&\sum_{f=0}^{n-1}2^{2f}\Big[C_{a(f),a(f)}(\tau)+C_{a(n+f),a(n+f)}(\tau)\Big]+ \\
&\sum_{0\le f<h\le n-1}2^{f+h}\Big\{(-1)^{c_f+c_h}\Big[C_{a(f),a(h)}(\tau)+C_{a(h),a(f)}(\tau)\Big] \\
&+(-1)^{d_f+d_h}\Big[C_{a(n+f),a(n+h)}(\tau)+C_{a(n+h),a(n+f)}(\tau)\Big]\Big\} \\
&+j\sum_{f=0}^{n-1}\sum_{h=0}^{n-1}(-1)^{c_f+d_h}2^{f+h}\Big[-C_{a(f),a(n+h)}(\tau)+C_{a(n+h),a(f)}(\tau)\Big],
\end{aligned}
$$

$$(88)$$

where $l = i + \tau$, and

$$\frac{1}{\lambda^2} C_{B,B}(\tau) = \sum_{f=0}^{n-1} 2^{2f} \left[C_{b^{(f)},b^{(f)}}(\tau) + C_{b^{(n+f)},b^{(n+f)}}(\tau) \right] +$$
$$\sum_{0 \le f < h \le n-1} 2^{f+h} \left\{ (-1)^{c_f + c_h} \left[C_{b^{(f)},b^{(h)}}(\tau) + C_{b^{(h)},b^{(f)}}(\tau) \right] \right.$$
$$\left. + (-1)^{d_f + d_h} \left[C_{b^{(n+f)},b^{(n+h)}}(\tau) + C_{b^{(n+h)},b^{(n+f)}}(\tau) \right] \right\}$$
$$+ j \sum_{f=0}^{n-1} \sum_{h=0}^{n-1} (-1)^{c_f + d_h} 2^{f+h} \left[- C_{b^{(f)},b^{(n+h)}}(\tau) + C_{b^{(n+h)},b^{(f)}}(\tau) \right]. \tag{89}$$

According to Theorem 5.1, the sequence pairs $(\underline{a}^{(f)}, \underline{b}^{(f)})$ and $(\underline{a}^{(n+f)}, \underline{b}^{(n+f)})$ $(0 \le f \le n - 1)$ form Golay CSPs. Therefore, for arbitrarily given $\tau > 0$ if the following equation system:

$$\begin{cases} C_{a^{(f)},a^{(h)}}(\tau) + C_{a^{(h)},a^{(f)}}(\tau) + C_{b^{(f)},b^{(h)}}(\tau) + C_{b^{(h)},b^{(f)}}(\tau) = 0 \\ \qquad\qquad (0 \le f < h \le n - 1) \tag{90a} \\[2mm] C_{a^{(n+f)},a^{(n+h)}}(\tau) + C_{a^{(n+h)},a^{(n+f)}}(\tau) + C_{b^{(n+f)},b^{(n+h)}}(\tau) + C_{b^{(n+h)},b^{(n+f)}}(\tau) = 0 \\ \qquad\qquad (0 \le f < h \le n - 1) \tag{90b} \\[2mm] -C_{a^{(f)},a^{(n+h)}}(\tau) + C_{a^{(n+h)},a^{(f)}}(\tau) - C_{b^{(f)},b^{(n+h)}}(\tau) + C_{b^{(n+h)},b^{(f)}}(\tau) = 0 \\ \qquad\qquad (0 \le f, h \le n - 1) \tag{90c} \end{cases}$$

holds, we must have

$$C_{A,A}(\tau) + C_{B,B}(\tau) = 0 \quad (\forall \, \tau > 0). \tag{91}$$

Hence, the cure to prove this theorem is to derive (90), which is investigated by the following three cases.

(i) Case I.
Case 1. On the equation (90a).

Denote the left-hand side of (90a) as $Lhs1(\tau)$. Thus, $Lhs1(\tau)$ is equivalent to

$$Lhs1(\tau) =$$
$$\sum_{i=0}^{N-\tau-1} \left[(-1)^{a^{(f)}(i)+a^{(h)}(l)} + (-1)^{b^{(f)}(i)+b^{(h)}(l)} + (-1)^{a^{(h)}(i)+a^{(f)}(l)} + (-1)^{b^{(h)}(i)+b^{(f)}(l)} \right]$$
$$= \sum_{i=0}^{N-\tau-1} \left[(-1)^{a(i)+s^{(f)}(i)+a(l)+s^{(h)}(l)} + (-1)^{a(i)+s^{(f)}(i)+\mu(i)+a(l)+s^{(h)}(l)+\mu(l)} \right.$$
$$\left. + (-1)^{a(i)+s^{(h)}(i)+a(l)+s^{(f)}(l)} + (-1)^{a(i)+s^{(h)}(i)+\mu(i)+a(l)+s^{(f)}(l)+\mu(l)} \right]$$
$$= \sum_{i=0}^{N-\tau-1} \left\{ (-1)^{a(i)+a(l)}(-1)^{s^{(f)}(i)+s^{(h)}(l)} \left[1 + (-1)^{\mu(i)+\mu(l)} \right] + \right.$$
$$\left. (-1)^{a(i)+a(l)}(-1)^{s^{(h)}(i)+s^{(f)}(l)} \left[1 + (-1)^{\mu(i)+\mu(l)} \right] \right\}$$
$$= \sum_{i=0}^{N-\tau-1} (-1)^{a(i)+a(l)} \left[(-1)^{s^{(f)}(i)+s^{(h)}(l)} + (-1)^{s^{(h)}(i)+s^{(f)}(l)} \right] \left[1 + (-1)^{\mu(i)+\mu(l)} \right].$$

$$(92)$$

Apparently, $Lhs1(\tau) = 0$ whenever $\mu_i \neq \mu_{i+\tau}$. When $\mu_i = \mu_{i+\tau}$, by employing Theorem 3.2, it is sufficient to check up that a pair of indices (i, i'), defined by Theorem 3.2, satisfies

$$(-1)^{s^{(f)}(i)+s^{(h)}(l)} + (-1)^{s^{(h)}(i)+s^{(f)}(l)} = (-1)^{s^{(f)}(i')+s^{(h)}(l')} + (-1)^{s^{(h)}(i')+s^{(f)}(l')},$$
$$(93)$$

where $l' = i' + \tau$.

Similar to Theorem 6.1, we also consider two cases: $w \geq v$ and $w \leq v - 1$, respectively.

(1): $w \geq v$.

Apparently, for $\forall r \in [0, 2n - 1]$ we have

$$\begin{cases} d_{r0} + d_{r1}i_{\sigma(w)} + d_{r2}i_{\sigma(w+1)} = d_{r0} + d_{r1}i'_{\sigma(w)} + d_{r2}i'_{\sigma(w+1)} \\ d_{r0} + d_{r1}l_{\sigma(w)} + d_{r2}l_{\sigma(w+1)} = d_{r0} + d_{r1}l'_{\sigma(w)} + d_{r2}l'_{\sigma(w+1)}, \end{cases}$$
$$(94)$$

which gives rise to

$$\begin{cases} s^{(f)}(i) = s^{(f)}(i') \\ s^{(h)}(i) = s^{(h)}(i') \\ s^{(f)}(l) = s^{(f)}(l') \\ s^{(h)}(l) = s^{(h)}(l'). \end{cases}$$
$$(95)$$

Consequently, (93) holds.

(2): $w \leq v - 1$.

Obviously, for $\forall\, r \in [0, 2n-1]$ we have

$$s^{(r)}(i) + s^{(r)}(l') = 2d_{r0} + d_{r1}\Big[i_{\sigma(w)} + l'_{\sigma(w)}\Big] + d_{r2}\Big[i_{\sigma(w+1)} + l'_{\sigma(w+1)}\Big]$$
$$= 2d_{r0} + d_{r1} + d_{r2}.$$

(96)

and

$$s^{(r)}(l) + s^{(r)}(i') = 2d_{r0} + d_{r1} + d_{r2}.$$

(97)

Thus, for $0 \le f < h \le n-1$, we have

$$\begin{cases} s^{(f)}(i) + s^{(f)}(l') + s^{(h)}(i') + s^{(h)}(l) = 2d_{f0} + 2d_{h0} + d_{f1} + d_{f2} + d_{h1} + d_{h2} \\ s^{(f)}(l) + s^{(f)}(i') + s^{(h)}(i) + s^{(h)}(l') = 2d_{f0} + 2d_{h0} + d_{f1} + d_{f2} + d_{h1} + d_{h2}. \end{cases}$$

(98)

Apparently, we have

$$\begin{cases} (-1)^{s^{(f)}(i)+s^{(h)}(l)} = (-1)^{s^{(f)}(l')+s^{(h)}(i')} \\ (-1)^{s^{(f)}(l)+s^{(h)}(i)} = (-1)^{s^{(f)}(i')+s^{(h)}(l')}, \end{cases}$$

(99)

when

$$d_{f1} + d_{f2} + d_{h1} + d_{h2} \equiv 0 \pmod{2}.$$

(100)

As a result, a sufficient condition that (90a) holds is

$$d_{f1} + d_{f2} + d_{h1} + d_{h2} \equiv 0 \pmod{2}\ (0 \le f < h \le n-1).$$

(101)

Case 2. On the equation (90b).

Write the left-hand side of (90b) as $Lhs2(\tau)$. Then, $Lhs2(\tau)$ is reduced to

$$Lhs2(\tau) = \sum_{i=0}^{N-\tau-1} (-1)^{a(i)+a(l)}\Big[(-1)^{s^{(n+f)}(i)+s^{(n+h)}(l)} + (-1)^{s^{(n+h)}(i)+s^{(n+f)}(l)}\Big]\Big[1 + (-1)^{\mu(i)+\mu(l)}\Big].$$

(102)

Similar to Case 1, a sufficient condition that (90b) holds is given by

$$d_{(n+f)1} + d_{(n+f)2} + d_{(n+h)1} + d_{(n+h)2} \equiv 0 \pmod{2}$$
$$(0 \le f < h \le n-1).$$

(103)

Case 3. On the equation (90c).

Denote the left-hand side of (90c) as $Lhs3(\tau)$. Then, $Lhs3(\tau)$ is

$$Lhs3(\tau) =$$
$$\sum_{i=0}^{N-\tau-1} (-1)^{a(i)+a(l)} \left[-(-1)^{s^{(f)}(i)+s^{(n+h)}(l)} + (-1)^{s^{(n+h)}(i)+s^{(f)}(l)} \right] \left[1 + (-1)^{\mu(i)+\mu(l)} \right],$$

$$(104)$$

where $0 \leq f, h \leq n-1$.

From Theorem 3.2, (90c) holds whenever

$$-(-1)^{s^{(f)}(i)+s^{(n+h)}(l)} + (-1)^{s^{(n+h)}(i)+s^{(f)}(l)} = -(-1)^{s^{(f)}(i')+s^{(n+h)}(l')} + (-1)^{s^{(n+h)}(i')+s^{(f)}(l')}.$$

$$(105)$$

Similarly, (105) is natural for $w \geq v$. For $w \leq v - 1$, due to

$$\begin{cases} s^{(f)}(i) + s^{(f)}(l') + s^{(n+h)}(i') + s^{(n+h)}(l) = \\ \quad 2d_{f0} + 2d_{(n+h)0} + d_{f1} + d_{f2} + d_{(n+h)1} + d_{(n+h)2} \\ s^{(f)}(l) + s^{(f)}(i') + s^{(n+h)}(i) + s^{(n+h)}(l') = \\ \quad 2d_{f0} + 2d_{(n+h)0} + d_{f1} + d_{f2} + d_{(n+h)1} + d_{(n+h)2}, \end{cases} \quad (106)$$

when

$$d_{f1} + d_{f2} + d_{(n+h)1} + d_{(n+h)2} \equiv 1 \pmod{2}, \tag{107}$$

we have

$$\begin{cases} (-1)^{s^{(f)}(i)+s^{(n+h)}(l)} = -(-1)^{s^{(n+h)}(i')+s^{(f)}(l')} \\ (-1)^{s^{(f)}(l)+s^{(n+h)}(l)} = -(-1)^{s^{(f)}(i')+s^{(n+h)}(l')}, \end{cases} \quad (108)$$

which produces (105).

As a consequence, a sufficient condition that (90c) holds is

$$d_{f1} + d_{f2} + d_{(n+h)1} + d_{(n+h)2} \equiv 1 \pmod{2}$$
$$(0 \leq f, h \leq n-1). \tag{109}$$

Combining Cases 1-3, a sufficient condition that the equation system (90) holds is

$$\begin{cases} d_{f1} + d_{f2} + d_{h1} + d_{h2} \equiv 0 \pmod{2} \\ \quad\quad (0 \leq f < h \leq n-1) \tag{110a} \\ \\ d_{(n+f)1} + d_{(n+f)2} + d_{(n+h)1} + d_{(n+h)2} \equiv 0 \pmod{2} \\ \quad\quad (0 \leq f < h \leq n-1) \tag{110b} \\ \\ d_{f1} + d_{f2} + d_{(n+h)1} + d_{(n+h)2} \equiv 1 \pmod{2} \\ \quad\quad (0 \leq f, h \leq n-1), \tag{110c} \end{cases}$$

Notice that for $1 \leq s < t \leq n - 1$,

$$\begin{cases} d_{01} + d_{02} + d_{s1} + d_{s2} \equiv 0 \quad (\text{mod } 2) \\ d_{01} + d_{02} + d_{t1} + d_{t2} \equiv 0 \quad (\text{mod } 2) \end{cases} \tag{111}$$

$$\stackrel{+}{\Longrightarrow} \quad d_{s1} + d_{s2} + d_{t1} + d_{t2} \equiv 0 \quad (\text{mod } 2), \tag{112}$$

for $0 \leq s < t \leq n - 1$,

$$\begin{cases} d_{01} + d_{02} + d_{(n+s)1} + d_{(n+s)2} \equiv 1 \quad (\text{mod } 2) \\ d_{01} + d_{02} + d_{(n+t)1} + d_{(n+t)2} \equiv 1 \quad (\text{mod } 2) \end{cases} \tag{113}$$

$$\stackrel{+}{\Longrightarrow} \quad d_{(n+s)1} + d_{(n+s)2} + d_{(n+t)1} + d_{(n+t)2} \equiv 0 \quad (\text{mod } 2), \tag{114}$$

and for $1 \leq s \leq n - 1$ and $0 \leq t \leq n - 1$,

$$\begin{cases} d_{01} + d_{02} + d_{s1} + d_{s2} \equiv 0 \quad (\text{mod } 2) \\ d_{01} + d_{02} + d_{(n+t)1} + d_{(n+t)2} \equiv 1 \quad (\text{mod } 2) \end{cases} \tag{115}$$

$$\stackrel{+}{\Longrightarrow} \quad d_{s1} + d_{s2} + d_{(n+t)1} + d_{(n+t)2} \equiv 1 \quad (\text{mod } 2). \tag{116}$$

Henceforth, (110) is compatible with

$$\begin{cases} d_{01} + d_{02} + d_{h1} + d_{h2} \equiv 0 \quad (\text{mod } 2) \ (1 \leq h \leq n - 1) \\ d_{01} + d_{02} + d_{(n+h)1} + d_{(n+h)2} \equiv 1 \ (\text{mod } 2) \ (0 \leq h \leq n - 1), \end{cases} \tag{117}$$

which completes our derivation for Case I.

Cases II and III.

These two cases are a special case of Case I with $d_{r2} = 0 \ (\forall r)$ and $w = 1$ or m. Employing the same derivation skills as ones in Theorem 6.1, Cases II and III hold.$\square$

Theorem 6.6. ([15])When $s^{(0)}(\underline{x}) = 0$, the number of 2^{2n}-QAM GCSs of length 2^m constructed from Theorems 6.5 is

$$\left[(m + 1)4^{2n-1} - m2^{2n-1} \right] \times (m!/2)2^{m+1} \ (m \geq 2, n \geq 2). \tag{118}$$

Proof. Quite a similar derivation to Theorem 6.2 needs to be done. In order to avoid too many tautologies, only some key equations are given below.

Due to $s^{(0)}(\underline{x}) = 0$, (84) degenerates to

$$\begin{cases} d_{h1} + d_{h2} \equiv 0 \ (\text{mod } 2) \ (1 \leq h \leq n - 1) \\ d_{(n+h)1} + d_{(n+h)2} \equiv 1 \ (\text{mod } 2) \ (0 \leq h \leq n - 1). \end{cases} \tag{119}$$

Let
$$S_I = \{d_0^{(r)} + d_1^{(r)} x_{\pi(1)} | 1 \leq r \leq 2n - 1\}, \tag{120}$$

$$S_{II} = \{d_0^{(r)} + d_1^{(r)} x_{\pi(m)} | 1 \leq r \leq 2n - 1\}, \tag{121}$$

and

$$S_{III} = \{d_0^{(r)} + d_1^{(r)} x_{\pi(w)} + d_2^{(r)} x_{\pi(w+1)} | 1 \leq r \leq 2n - 1, 1 \leq w \leq m - 1,$$

with

$$d_1^{(h)} + d_2^{(h)} \equiv 0 \pmod{2} \ (1 \leq h \leq n - 1) \tag{122a}$$

and

$$d_1^{(n+h)} + d_2^{(n+h)} \equiv 1 \pmod{2} \ (0 \leq h \leq n - 1)\}, \tag{122b}$$
$$\tag{122}$$

where $d_l^{(r)} \in Z_2$ ($1 \leq r \leq 2n - 1$ and $0 \leq l \leq 2$).

Similarly, the number of these 2^{2n}-QAM GCSs equals

$$|S_I \bigcup S_{II} \bigcup S_{III}| \times (m!/2) 2^{m+1},$$

where

$$\begin{aligned}
&|S_I| = |S_{II}| = 4^{2n-1}, \\
&|S_{III}| = (m - 1) 4^{2n-1} - (m - 2) 2^{2n-1}, \\
&|S_{I,II}| = |S_I \bigcap S_{II}| = |\{d_0^{(r)} | 1 \leq r \leq 2n - 1\}| = 2^{2n-1}, \\
&|S_{I,III}| = |S_I \bigcap S_{III}| = |\{d_0^{(r_1)}, d_0^{(n+r_2)} + x_{\pi(1)}| \\
&\qquad\qquad 1 \leq r_1 \leq n - 1, 0 \leq r_2 \leq n - 1\}| = 2^{2n-1}, \\
&|S_{II,III}| = |S_{II} \bigcap S_{III}| = |\{d_0^{(r_1)}, d_0^{(n+r_2)} + x_{\pi(m)}| \\
&\qquad\qquad 1 \leq r_1 \leq n - 1, 0 \leq r_2 \leq n - 1\}| = 2^{2n-1}, \\
&|S_{I,II,III}| = |S_I \bigcap S_{II} \bigcap S_{III}| = |\{d_0^{(r)} | 1 \leq r \leq 2n - 1\}| = 2^{2n-1}.
\end{aligned}$$

We complete the proof. $\square$

Theorem 6.7. Let $\underline{A}$ and $\underline{B}$ be a 2^{2n}-QAM GCS from Theorems 6.5 with length $N = 2^m$. Then, the PEP that is associated with the sequence $\underline{A}$ is bounded by

$$\mathrm{PEP}(\underline{A}) \leq \frac{6N(2^n - 1)}{2^n + 1}. \tag{123}$$

Proof. Similar to Theorem 6.4, we have

$$\mathrm{PEP}(\underline{A}) \leq 2 \sum_{i=0}^{N-1} |A_i|^2. \tag{124}$$

Since

$$\frac{1}{\lambda^2}|A_i|^2 =$$

$$\left|\left[\sum_{f=0}^{n-1}(-1)^{c_f}2^f(-1)^{a_i^{(f)}} + j\sum_{g=0}^{n-1}(-1)^{d_g}2^g(-1)^{a_i^{(n+g)}}\right]\right|^2$$

$$= \left|\sum_{f=0}^{n-1}(-1)^{c_f}2^f(-1)^{a_i^{(f)}}\right|^2 + \left|\sum_{p=0}^{n-1}(-1)^{d_g}2^g(-1)^{a_i^{(n+g)}}\right|^2 \qquad (125)$$

$$\leq 2(2^n - 1)^2,$$

this theorem follows immediately.□

In practical applications, implementation of QAM constellation of larger size is a great challenge. For this reason, 16-QAM, 64-QAM, and 256-QAM constellations are preferred. In what follows, resultant 16- and 64- QAM GCSs in Theorem 6.5 will be investigated in detail, including upper bounds of their PEPs, and only all solutions of (84) are given in 256-QAM GCSs.

6.3.1. 16-QAM GCSs in Theorem 6.5

(84) degenerates to

$$\begin{cases} d_{01} + d_{02} + d_{11} + d_{12} \equiv 0 \ (\text{mod } 2) & (126a) \\ d_{01} + d_{02} + d_{21} + d_{22} \equiv 1 \ (\text{mod } 2) & (126b) \\ d_{01} + d_{02} + d_{31} + d_{32} \equiv 1 \ (\text{mod } 2). & (126c) \end{cases}$$

By a computer, all solutions $d_{01}d_{02}d_{11}d_{12}d_{21}d_{22}d_{31}d_{32}$'s (total number 32) in (126) are given below.

<u>00000101</u> 00000110 00001001 <u>00001010</u> 00110101 00110110 00111001
00111010 <u>01010000</u> 01010011 01011100 01011111 01100000 01100011
01101100 01101111 10010000 10010011 10011100 10011111 <u>10100000</u>
10100011 10101100 10101111 11000101 11000110 11001001 11001010
  11110101 11110110 11111001 11111010.

It should be noted that four solutions with underlines result in overlapping offsets. For example, the offsets from 00000101 and 00001010 are

$$\begin{cases} s^{(0)}(\underline{x}) = d_{00} \\ s^{(1)}(\underline{x}) = d_{10} \\ s^{(2)}(\underline{x}) = d_{20} + x_{\sigma(w+1)} \ (1 \leq w \leq m-1) \\ s^{(3)}(\underline{x}) = d_{30} + x_{\sigma(w+1)} \ (1 \leq w \leq m-1) \end{cases}$$

and

$$\begin{cases} s^{(0)}(\underline{x}) = d_{00} \\ s^{(1)}(\underline{x}) = d_{10} \\ s^{(2)}(\underline{x}) = d_{20} + x_{\sigma(w)} \ (1 \le w \le m) \\ s^{(3)}(\underline{x}) = d_{30} + x_{\sigma(w)} \ (1 \le w \le m), \end{cases}$$

respectively. Apparently, when the variable w in the former ranges the range from 1 to $m-1$, and the variable w in the latter takes on each value over the range from 2 to m, the same offsets appear. All distinct solutions and upper bounds of PEPs of resultant 16-QAM GCSs in Case I are listed in Table 2, and Table 3 for Cases II and III, where $|d_{10} - d_{00}| = a$ and $|d_{30} - d_{20}| = b$. Notice that the coefficients d_{r0}'s $(0 \le r \le 3)$ are pairwise independent. Hence, the total number of the solution of all the coefficients in (84) is $30 \times 16 = 480$.

Consider Case 3 in Table 3 and Case II, that is, $(d_{01}, d_{11}, d_{21}, d_{31}) = (0, 0, 1, 0)$. Then, we have

$$10|A_i|^2 = \left[1 + 2(-1)^a\right]^2 + \left[1 + 2(-1)^b(-1)^{i_{\sigma(1)}}\right]^2$$
$$= \begin{cases} \left|1 + 2(-1)^a\right|^2 + \left|1 + 2(-1)^b\right|^2 & \text{if } i_{\sigma(1)} = 0 \\ \left|1 + 2(-1)^a\right|^2 + \left|1 - 2(-1)^b\right|^2 & \text{if } i_{\sigma(1)} = 1. \end{cases}$$

Consider $(a, b) = (0, 0)$. Notice that $|B_i|^2 = |A_i|^2 \ (0 \le i < 2^m)$. Therefore, we have

$$C_{A,A}(0) + C_{B,B}(0) = \sum_{i=0}^{2^m - 1} [|A_i|^2 + |B_i|^2] = \frac{2}{10}(18m_0 + 10m_1),$$

where $m_0 = |\{i|i_{\sigma(1)} = 0, 0 \le i < 2^m\}|$ and $m_1 = |\{i|i_{\sigma(1)} = 1, 0 \le i < 2^m\}|$.

Further, due to $|\{i|i_{\sigma(1)} = 0, 0 \le i < 2^m\}| = |\{i|i_{\sigma(1)} = 1, 0 \le i < 2^m\}| = N/2$, we have

$$C_{A,A}(0) + C_{B,B}(0) = \frac{28}{10}N = 2.8N,$$

which implies that for the proposed 16-QAM GCSs in this case, the upper bound of PEP($\underline{A}$) is $2.8N$.

Table 2. All possible offsets and PEP upper bounds of the resultant 16-QAM GCSs in Case I

case	$d_{01}d_{02}$	$d_{11}d_{22}$	$d_{21}d_{22}$	$d_{31}d_{32}$	w	bounds under (a,b)			
						(0,0)	(0,1)	(1,0)	(1,1)
1	00	00	01	10	$1 \leq w < m$	2.8N	2.8N	1.2N	1.2N
2	00	00	10	01	$1 \leq w < m$	2.8N	2.8N	1.2N	1.2N
3	00	00	10	10	$1 \leq w \leq m$	3.6N	2N	2N	0.4N
4	00	11	01	01	$1 \leq w < m$	2.8N	1.2N	2.8N	1.2N
5	00	11	01	10	$1 \leq w < m$	2N	2N	2N	2N
6	00	11	10	01	$1 \leq w < m$	2N	2N	2N	2N
7	00	11	10	10	$1 \leq w < m$	2.8N	1.2N	2.8N	N1.2
8	01	01	00	11	$1 \leq w < m$	2.8N	2.8N	1.2N	1.2N
9	01	01	11	00	$1 \leq w < m$	2.8N	2.8N	1.2N	1.2N
10	01	01	11	11	$1 \leq w < m$	3.6N	2N	2N	0.4N
11	01	10	00	00	$1 \leq w < m$	2.8N	1.2N	2.8N	1.2N
12	01	10	00	11	$1 \leq w < m$	2N	2N	2N	2N
13	01	10	11	00	$1 \leq w < m$	2N	2N	2N	2N
14	01	10	11	11	$1 \leq w < m$	2.8N	1.2N	2.8N	1.2N
15	10	01	00	00	$1 \leq w < m$	2.8N	1.2N	2.8N	1.2N
16	10	01	00	11	$1 \leq w < m$	2N	2N	2N	2N
17	10	01	11	00	$1 \leq w < m$	2N	2N	2N	2N
18	10	01	11	11	$1 \leq w < m$	2.8N	1.2N	2.8N	1.2N
19	10	10	00	00	$1 \leq w \leq m$	3.6N	2N	2N	0.4N
20	10	10	00	11	$1 \leq w < m$	2.8N	2.8N	1.2N	1.2N
21	10	10	11	00	$1 \leq w < m$	2.8N	2.8N	1.2N	1.2N
22	10	10	11	11	$1 \leq w < m$	3.6N	2N	2N	0.4N
23	11	00	01	01	$1 \leq w < m$	2.8N	1.2N	2.8N	1.2N
24	11	00	01	10	$1 \leq w < m$	2N	2N	2N	2N
25	11	00	10	01	$1 \leq w < m$	2N	2N	2N	2N
26	11	00	10	10	$1 \leq w < m$	2.8N	1.2N	2.8N	1.2N
27	11	11	01	01	$1 \leq w < m$	3.6N	2N	2N	0.4N
28	11	11	01	10	$1 \leq w < m$	2.8N	2.8N	1.2N	1.2N
29	11	11	10	01	$1 \leq w < m$	2.8N	2.8N	1.2N	1.2N
30	11	11	10	10	$1 \leq w < m$	3.6N	2N	2N	0.4N

Example 6.2. Here is a numerical example to help the reader understand. Set $m = 4$, $\sigma(i) = i$ $(1 \leq i \leq 4)$, and $a(\underline{x}) = \sum_{i=1}^{3} x_i x_{i+1}$, and consider Case 13 in Table 3 in Case II under $(d_{00}, d_{10}, d_{20}, d_{30}) = (0, 1, 0, 1)$, namely, $(a, b) =$

Table 3. All possible offsets and PEP upper bounds of the resultant 16-QAM GCSs in Cases II and III

case	d_{01}	d_{11}	d_{21}	d_{31}	w	bounds under (ab)			
						(00)	(01)	(10)	(11)
1	0	0	0	0	$1/m$	3.6	2	2	0.4
2	0	0	0	1	$1/m$	2.8	2.8	1.2	1.2
3	0	0	1	0	$1/m$	2.8	2.8	1.2	1.2
4	0	0	1	1	$1/m$	3.6	2	2	0.4
5	0	1	0	0	$1/m$	2.8	1.2	2.8	1.2
6	0	1	0	1	$1/m$	2	2	2	2
7	0	1	1	0	$1/m$	2	2	2	2
8	0	1	1	1	$1/m$	2.8	1.2	2.8	1.2
9	1	0	0	0	$1/m$	2.8	1.2	2.8	1.2
10	1	0	0	1	$1/m$	2	2	2	2
11	1	0	1	0	$1/m$	2	2	2	2
12	1	0	1	1	$1/m$	2.8	1.2	2.8	1.2
13	1	1	0	0	$1/m$	3.6	2	2	0.4
14	1	1	0	1	$1/m$	2.8	2.8	1.2	1.2
15	1	1	1	0	$1/m$	2.8	2.8	1.2	1.2
16	1	1	1	1	$1/m$	3.6	2	2	0.4

$(1, 1)$. More clearly, we have

$$\begin{cases} s^{(0)}(\underline{x}) = 1 + x_1 \\ s^{(1)}(\underline{x}) = x_1 \\ s^{(2)}(\underline{x}) = 0 \\ s^{(3)}(\underline{x}) = 1 \end{cases}$$

and $\mu(\underline{x}) = x_4$, then the resultant 16-QAM GCSs of length $N = 16$ are
$\underline{A} = \frac{1}{\sqrt{10}}[1 - j, 1 - j, 1 - j, -1 + j, 1 - j, 1 - j, -1 + j, 1 - j, -1 - j, -1 - j, -1 - j, 1 + j, 1 + j, 1 + j, -1 - j, 1 + j]$
$\underline{B} = \frac{1}{\sqrt{10}}[1 - j, -1 + j, 1 - j, 1 - j, 1 - j, -1 + j, -1 + j, -1 + j, -1 - j, 1 + j, -1 - j, -1 - j, 1 + j, -1 - j, -1 - j, -1 - j]$,
the sum of whose autocorrelation functions with time shift $\tau \in [0, 15]$ is

$$C_{A,A}(\tau) + C_{B,B}(\tau) = (6.4, 0, 0, 0, 0, 0, 0, 0, 0, 0, 0, 0, 0, 0, 0, 0).$$

Hence, the PEP that is associated with the sequence $\underline{A}$ satisfies

$$\text{PEP}(\underline{A}) \leq 6.4 = 0.4N.$$

which is exactly such a value given in Table 3.

6.3.2. 64-QAM GCSs in Theorem 6.5

(84) is reduced to

$$\begin{cases} d_{01} + d_{02} + d_{11} + d_{12} \equiv 0 \ (\text{mod } 2) \\ d_{01} + d_{02} + d_{21} + d_{22} \equiv 0 \ (\text{mod } 2) \\ d_{01} + d_{02} + d_{31} + d_{32} \equiv 1 \ (\text{mod } 2) \\ d_{01} + d_{02} + d_{41} + d_{42} \equiv 1 \ (\text{mod } 2) \\ d_{01} + d_{02} + d_{51} + d_{52} \equiv 1 \ (\text{mod } 2), \end{cases} \quad (127)$$

whose all the solutions $d_{01}d_{02}d_{11}d_{12}\ d_{31}d_{32}d_{41}d_{42}d_{51}d_{52}$'s (total number 128) are given by

<u>000000010101</u>   000000010110   000000011001
000000011010   000000100101   000000100110
000000101001   <u>000000101010</u>   000011010101
000011010110   000011011001   000011011010
000011100101   000011100110   000011101001
000011101010   001100010101   001100010110
001100011001   001100011010   001100100101
001100100110   001100101001   001100101010
001111010101   001111010110   001111011001
001111011010   001111100101   001111100110
001111101001   001111101010   <u>010101000000</u>
010101000011   010101001100   010101001111
010101110000   010101110011   010101111100
010101111111   010110000000   010110000011
010110001100   010110001111   010110110000
010110110011   010110111100   010110111111
011001000000   011001000011   011001001100
011001001111   011001110000   011001110011
011001111100   011001111111   011010000000
011010000011   011010001100   011010001111
011010110000   011010110011   011010111100
011010111111   100101000000   100101000011
100101001100   100101001111   100101110000
100101110011   100101111100   100101111111
100110000000   100110000011   100110001100
100110001111   100110110000   100110110011

$$100110111100 \quad 100110111111 \quad 101001000000$$
$$101001000011 \quad 101001001100 \quad 101001001111$$
$$101001110000 \quad 101001110011 \quad 101001111100$$
$$101001111111 \quad \underline{101010000000} \quad 101010000011$$
$$101010001100 \quad 101010001111 \quad 101010110000$$
$$101010110011 \quad 101010111100 \quad 101010111111$$
$$110000010101 \quad 110000010110 \quad 110000011001$$
$$110000011010 \quad 110000100101 \quad 110000100110$$
$$110000101001 \quad 110000101010 \quad 110011010101$$
$$110011010110 \quad 110011011001 \quad 110011011010$$
$$110011100101 \quad 110011100110 \quad 110011101001$$
$$110011101010 \quad 111100010101 \quad 111100010110$$
$$111100011001 \quad 111100011010 \quad 111100100101$$
$$111100100110 \quad 111100101001 \quad 111100101010$$
$$111111010101 \quad 111111010110 \quad 111111011001$$
$$111111011010 \quad 111111100101 \quad 111111100110$$
$$111111101001 \quad 111111101010.$$

After removing all the overlaps, we obtain 126 distinct solutions. Notice that the coefficients d_{r0}'s ($0 \leq r \leq 5$) independently ranges over the range from 0 to 1. Therefore, we have 8064 ($=2^6 \times 126$) distinct offsets.

Denote the vector $(a, b, c, d) = (|d_{10} - d_{00}|, |d_{20} - d_{00}|, |d_{40} - d_{30}|, |d_{50} - d_{30}|)$. These vectors vary from $(0,0,0,0)$ to $(1,1,1,1)$ owing to pairwise independent coefficients d_{r0}'s ($0 \leq r \leq 5$).

Consider $(d_{01}d_{02}, d_{11}d_{22}, d_{21}d_{22}, d_{31}d_{32}, d_{41}d_{42}, d_{51}d_{52}) = (00, 00, 11, 10, 01, 01)$ and $(c_0, c_1, c_2, d_0, d_1, d_2) = (0, 0, 0, 0, 0, 0)$. Then, (87) is equivalent to

$$\begin{cases} \sqrt{42}A_i = \\ \quad \left[(-1)^{d_{00}+a(i)} + 2(-1)^{d_{10}+a(i)} + 4(-1)^{d_{20}+i_\sigma(w)+i_\sigma(w+1)+a(i)} \right] \\ \quad + j \left[(-1)^{d_{30}+i_\sigma(w)+a(i)} + 2(-1)^{d_{40}+i_\sigma(w+1)+a(i)} + \right. \\ \qquad\qquad\qquad\qquad \left. 4(-1)^{d_{50}+i_\pi(w+1)+a_i} \right] \\ \sqrt{42}B_i = \\ \quad \left[(-1)^{d_{00}+b(i)} + 2(-1)^{d_{10}+b(i)} + 4(-1)^{d_{20}+i_\sigma(w)+i_\sigma(w+1)+b(i)} \right] \\ \quad + j \left[(-1)^{d_{30}+i_\sigma(w)+b(i)} + 2(-1)^{d_{40}+i_\sigma(w+1)+b(i)} + \right. \\ \qquad\qquad\qquad\qquad \left. 4(-1)^{d_{50}+i_\sigma(w+1)+b(i)} \right]. \end{cases}$$

Further, we have

$$
42|A_i|^2 = \left| (-1)^{a(i)} \left[1 + 2(-1)^a + 4(-1)^b(-1)^{i_{\sigma(w)}+i_{\sigma(w+1)}} \right] \right|^2
$$
$$
+ \left| (-1)^{i_{\sigma(w)}+a(i)} \left[1 + 2(-1)^c(-1)^{i_{\sigma(w+1)}-i_{\sigma(w)}} + \right. \right.
$$
$$
\left. \left. 4(-1)^d(-1)^{i_{\sigma(w+1)}-i_{\sigma(w)}} \right] \right|^2
$$
$$
= \begin{cases} \left| 1 + 2(-1)^a + 4(-1)^b \right|^2 + \left| 1 + 2(-1)^c + 4(-1)^d \right|^2 \\ \qquad\qquad\qquad \text{if } i_{\sigma(w+1)} = i_{\sigma(w)} \\ \left| 1 + 2(-1)^a - 4(-1)^b \right|^2 + \left| 1 - 2(-1)^c - 4(-1)^d \right|^2 \\ \qquad\qquad\qquad \text{if } i_{\sigma(w+1)} \neq i_{\sigma(w)}. \end{cases}
$$

Apparently, when $(a, b, c, d) = (0, 0, 0, 0)$, $|A_i|^2$ is reduced to

$$
42|A_i|^2 = \begin{cases} 7^2 + 7^2 = 98 & \text{if } i_{\sigma(w+1)} = i_{\sigma(w)} \\ 1 + 5^2 = 26 & \text{if } i_{\sigma(w+1)} \neq i_{\sigma(w)}. \end{cases}
$$

Due to $|B_i|^2 = |A_i|^2$ $(0 \leq i < 2^m)$, we have

$$
C_{A,A}(0) + C_{B,B}(0) = \frac{1}{42}(196m_0 + 52m_1),
$$

where $m_0 = |\{i | i_{\sigma(w+1)} = i_{\sigma(w)}, 0 \leq i < 2^m\}|$ and $m_1 = |\{i | i_{\sigma(w+1)} \neq i_{\sigma(w)}, 0 \leq i < 2^m\}|$.

Further, due to $|\{i | i_{\sigma(w+1)} = i_{\sigma(w)}, 0 \leq i < 2^m\}| = |\{i | i_{\sigma(w+1)} \neq i_{\sigma(w)}, 0 \leq i < 2^m\}| = N/2$, we have

$$
C_{A,A}(0) + C_{B,B}(0) = \frac{124}{42}N = 2.95N,
$$

which implies that for the proposed 64-QAM GCSs in this case, the upper bound of PEP($\underline{A}$) is $2.95N$.

Here are a numerical example to help the reader understand.

Example 6.3. We set $m = 5$, $\sigma(i) = i$ $(1 \leq i \leq 5)$, and $a(\underline{x}) = \sum_{i=1}^{4} x_i x_{i+1}$, and consider $w = 1$ and $(d_{00}, d_{10}, d_{20}, d_{30}, d_{40}, d_{50}) = (0, 0, 1, 0, 0, 0)$, that is,

$(a, b, c, d) = (0, 1, 0, 0)$. More clearly, we have

$$\begin{cases} s^{(0)}(\underline{x}) = 0 \\ s^{(1)}(\underline{x}) = 0 \\ s^{(2)}(\underline{x}) = 1 + x_1 + x_2 \\ s^{(3)}(\underline{x}) = x_2 \\ s^{(4)}(\underline{x}) = x_1 \\ s^{(5)}(\underline{x}) = x_1 \end{cases} \tag{128}$$

and $\mu(\underline{x}) = x_1$, then the resultant 64-QAM GCSs of length $N = 32$ is

$\underline{A} = \frac{1}{\sqrt{42}}[-1 + 7j, -1 + 7j, -1 + 7j, 1 - 7j, -1 + 7j, -1 + 7j, 1 - 7j, -1 + 7j, 7 + 5j, 7 + 5j, 7 + 5j, -7 - 5j, -7 - 5j, -7 - 5j, 7 + 5j, -7 - 5j, 7 - 5j, 7 - 5j, 7 - 5j, -7 + 5j, 7 - 5j, 7 - 5j, -7 + 5j, 7 - 5j, 1 + 7j, 1 + 7j, 1 + 7j, -1 - 7j, -1 - 7j, -1 - 7j, 1 + 7j, -1 - 7j]$

$\underline{B} = \frac{1}{\sqrt{42}}[-1 + 7j, 1 - 7j, -1 + 7j, -1 + 7j, -1 + 7j, 1 - 7j, 1 - 7j, 1 - 7j, 7 + 5j, -7 - 5j, 7 + 5j, 7 + 5j, -7 - 5j, 7 + 5j, 7 + 5j, 7 + 5j, 7 - 5j, -7 + 5j, 7 - 5j, 7 - 5j, 7 - 5j, -7 + 5j, -7 + 5j, -7 + 5j, 1 + 7j, -1 - 7j, 1 + 7j, 1 + 7j, -1 - 7j, 1 + 7j, 1 + 7j, 1 + 7j]$.

After calculation, we have

$$C_{A,A}(0) + C_{B,B}(0) = \frac{3968}{42} = \frac{124}{42}N = 2.95N.$$

Tables 4-14 illuminate the PEP upper bounds of 64-QAM CSPs in Cases I, II and III in Theorem 6.5.

6.3.3. 256-QAM GCSs in Theorem 6.5

(84) is reduced to

$$\begin{cases} d_{01} + d_{02} + d_{11} + d_{12} \equiv 0 \pmod 2 \\ d_{01} + d_{02} + d_{21} + d_{22} \equiv 0 \pmod 2 \\ d_{01} + d_{02} + d_{31} + d_{32} \equiv 0 \pmod 2 \\ d_{01} + d_{02} + d_{41} + d_{42} \equiv 1 \pmod 2 \\ d_{01} + d_{02} + d_{51} + d_{52} \equiv 1 \pmod 2 \\ d_{01} + d_{02} + d_{61} + d_{62} \equiv 1 \pmod 2 \\ d_{01} + d_{02} + d_{71} + d_{72} \equiv 1 \pmod 2, \end{cases} \tag{129}$$

whose all the solutions (total number 512) $d_{01}d_{02}d_{11}d_{12}$ $d_{31}d_{32}d_{41}d_{42}d_{51}d_{52}d_{61}d_{62}$ $d_{71}d_{72}$'s are given by

0000000001010101	0000000001010110	0000000001011001
0000000001011010	0000000001100101	0000000001100110
0000000001101001	0000000001101010	0000000010010101
0000000010010110	0000000010011001	0000000010011010
0000000010100101	0000000010100110	0000000010101001
0000000010101010	0000001101010101	0000001101010110
0000001101011001	0000001101011010	0000001101100101
0000001101100110	0000001101101001	0000001101101010
0000001110010101	0000001110010110	0000001110011001
0000001110011010	0000001110100101	0000001110100110
0000001110101001	0000001110101010	0000110001010101
0000110001010110	0000110001011001	0000110001011010
0000110001100101	0000110001100110	0000110001101001
0000110001101010	0000110010010101	0000110010010110
0000110010011001	0000110010011010	0000110010100101
0000110010100110	0000110010101001	0000110010101010
0000111101010101	0000111101010110	0000111101011001
0000111101011010	0000111101100101	0000111101100110
0000111101101001	0000111101101010	0000111110010101
0000111110010110	0000111110011001	0000111110011010
0000111110100101	0000111110100110	0000111110101001
0000111110101010	0011000001010101	0011000001010110
0011000001011001	0011000001011010	0011000001100101
0011000001100110	0011000001101001	0011000001101010
0011000010010101	0011000010010110	0011000010011001
0011000010011010	0011000010100101	0011000010100110
0011000010101001	0011000010101010	0011001101010101
0011001101010110	0011001101011001	0011001101011010
0011001101100101	0011001101100110	0011001101101001
0011001101101010	0011001110010101	0011001110010110
0011001110011001	0011001110011010	0011001110100101
0011001110100110	0011001110101001	0011001110101010
0011110001010101	0011110001010110	0011110001011001
0011110001011010	0011110001100101	0011110001100110
0011110001101001	0011110001101010	0011110010010101
0011110010010110	0011110010011001	0011110010011010
0011110010100101	0011110010100110	0011110010101001
0011110010101010	0011111101010101	0011111101010110
0011111101011001	0011111101011010	0011111101100101
0011111101100110	0011111101101001	0011111101101010
0011111110010101	0011111110010110	0011111110011001
0011111110011010	0011111110100101	0011111110100110
0011111110101001	0011111110101010	0101010100000000
0101010100000011	0101010100001100	0101010100001111
0101010100110000	0101010100110011	0101010100111100
0101010100111111	0101010111000000	0101010111000011
0101010111001100	0101010111001111	0101010111110000
0101010111110011	0101010111111100	0101010111111111
0101011000000000	0101011000000011	0101011000001100
0101011000001111	0101011000110011	0101011000110011

```
0101011011111111   0101100100000000   0101100100000011
0101100100001100   0101100100001111   0101100100110000
0101100100110011   0101100100111100   0101100100111111
0101100111000000   0101100111000011   0101100111001100
0101100111001111   0101100111110000   0101100111110011
0101100111111100   0101100111111111   0101101000000000
0101101000000011   0101101000001100   0101101000001111
0101101000110000   0101101000110011   0101101000111100
0101101000111111   0101101011000000   0101101011000011
0101101011001100   0101101011001111   0101101011110000
0101101011110011   0101101011111100   0101101011111111
0110010100000000   0110010100000011   0110010100001100
0110010100001111   0110010100110000   0110010100110011
0110010100111100   0110010100111111   0110010111000000
0110010111000011   0110010111001100   0110010111001111
0110010111110000   0110010111110011   0110010111111100
0110010111111111   0110011000000000   0110011000000011
0110011000001100   0110011000001111   0110011000110000
0110011000110011   0110011000111100   0110011000111111
0110011011000000   0110011011000011   0110011011001100
0110011011001111   0110011011110000   0110011011110011
0110011011111100   0110011011111111   0110100100000000
0110100100000011   0110100100001100   0110100100001111
0110100100110000   0110100100110011   0110100100111100
0110100100111111   0110100111000000   0110100111000011
0110100111001100   0110100111001111   0110100111110000
0110100111110011   0110100111111100   0110100111111111
0110101000000000   0110101000000011   0110101000001100
0110101000001111   0110101000110000   0110101000110011
0110101000111100   0110101000111111   0110101011000000
0110101011000011   0110101011001100   0110101011001111
0110101011110000   0110101011110011   0110101011111100
0110101011111111   1001010100000000   1001010100000011
1001010100001100   1001010100001111   1001010100110000
1001010100110011   1001010100111100   1001010100111111
1001010111000000   1001010111000011   1001010111001100
1001010111001111   1001010111110000   1001010111110011
1001010111111100   1001010111111111   1001011000000000
1001011000000011   1001011000001100   1001011000001111
1001011000110000   1001011000110011   1001011000111100
1001011000111111   1001011011000000   1001011011000011
1001011011001100   1001011011001111   1001011011110000
1001011011110011   1001011011111100   1001011011111111
1001100100000000   1001100100000011   1001100100001100
1001100100001111   1001100100110000   1001100100110011
1001100100111100   1001100100111111   1001100111000000
1001100111000011   1001100111001100   1001100111001111
1001100111110000   1001100111110011   1001100111111100
1001100111111111   1001101000000000   1001101000000011
1001101000001100   1001101000001111   1001101000110000
1001101000110011   1001101000111100   1001101000111111
1001101011000000   1001101011000011   1001101011001100
1001101011001111   1001101011110000   1001101011110011
```

```
1001101011111100   1001101011111111   1010010100000000
1010010100000011   1010010100001100   1010010100001111
1010010100110000   1010010100110011   1010010100111100
1010010100111111   1010010111000000   1010010111000011
1010010111001100   1010010111001111   1010010111110000
1010010111110011   1010010111111100   1010010111111111
1010011000000000   1010011000000011   1010011000001100
1010011000001111   1010011000110000   1010011000110011
1010011000111100   1010011000111111   1010011011000000
1010011011000011   1010011011001100   1010011011001111
1010011011110000   1010011011110011   1010011011111100
1010011011111111   1010100100000000   1010100100000011
1010100100001100   1010100100001111   1010100100110000
1010100100110011   1010100100111100   1010100100111111
1010100111000000   1010100111000011   1010100111001100
1010100111001111   1010100111110000   1010100111110011
1010100111111100   1010100111111111   1010101000000000
1010101000000011   1010101000001100   1010101000001111
1010101000110000   1010101000110011   1010101000111100
1010101000111111   1010101011000000   1010101011000011
1010101011001100   1010101011001111   1010101011110000
1010101011110011   1010101011111100   1010101011111111
1100000001010101   1100000001010110   1100000001011001
1100000001011010   1100000001100101   1100000001100110
1100000001101001   1100000001101010   1100000010010101
1100000010010110   1100000010011001   1100000010011010
1100000010100101   1100000010100110   1100000010101001
1100000010101010   1100001101010101   1100001101010110
1100001101011001   1100001101011010   1100001101100101
1100001101100110   1100001101101001   1100001101101010
1100001110010101   1100001110010110   1100001110011001
1100001110011010   1100001110100101   1100001110100110
1100001110101001   1100001110101010   1100110001010101
1100110001010110   1100110001011001   1100110001011010
1100110001100101   1100110001100110   1100110001101001
1100110001101010   1100110010010101   1100110010010110
1100110010011001   1100110010011010   1100110010100101
1100110010100110   1100110010101001   1100110010101010
1100111101010101   1100111101010110   1100111101011001
1100111101011010   1100111101100101   1100111101100110
1100111101101001   1100111101101010   1100111110010101
1100111110010110   1100111110011001   1100111110011010
1100111110100101   1100111110100110   1100111110101001
1100111110101010   1111000001010101   1111000001010110
1111000001011001   1111000001011010   1111000001100101
1111000001100110   1111000001101001   1111000001101010
1111000010010101   1111000010010110   1111000010011001
1111000010011010   1111000010100101   1111000010100110
1111000010101001   1111000010101010   1111001101010101
1111001101010110   1111001101011001   1111001101011010
1111001101100101   1111001101100110   1111001101101001
1111001101101010   1111001110010101   1111001110010110
1111001110011001   1111001110011010   1111001110100101
```

```
1 1 1 1 0 0 1 1 1 0 1 0 0 1 1 0   1 1 1 1 0 0 1 1 1 0 1 0 1 0 0 1   1 1 1 1 0 0 1 1 1 0 1 0 1 0 1 0
1 1 1 1 1 1 0 0 0 1 0 1 0 1 0 1   1 1 1 1 1 1 0 0 0 1 0 1 0 1 1 0   1 1 1 1 1 1 0 0 0 1 0 1 1 0 0 1
1 1 1 1 1 1 0 0 0 1 0 1 1 0 1 0   1 1 1 1 1 1 0 0 0 1 1 0 0 1 0 1   1 1 1 1 1 1 0 0 0 1 1 0 0 1 1 0
1 1 1 1 1 1 0 0 0 1 1 0 1 0 0 1   1 1 1 1 1 1 0 0 0 1 1 0 1 0 1 0   1 1 1 1 1 1 0 0 1 0 0 1 0 1 0 1
1 1 1 1 1 1 0 0 1 0 0 1 0 1 1 0   1 1 1 1 1 1 0 0 1 0 0 1 1 0 0 1   1 1 1 1 1 1 0 0 1 0 0 1 1 0 1 0
1 1 1 1 1 1 0 0 1 0 1 0 0 1 0 1   1 1 1 1 1 1 0 0 1 0 1 0 0 1 1 0   1 1 1 1 1 1 0 0 1 0 1 0 1 0 0 1
1 1 1 1 1 1 0 0 1 0 1 0 1 0 1 0   1 1 1 1 1 1 1 0 1 0 1 0 1 0 1   1 1 1 1 1 1 1 0 1 0 1 0 1 1 0
1 1 1 1 1 1 1 0 1 0 1 1 0 0 1   1 1 1 1 1 1 1 0 1 0 1 1 0 1 0   1 1 1 1 1 1 1 0 1 1 0 0 1 0 1
1 1 1 1 1 1 1 0 1 1 0 0 1 1 0   1 1 1 1 1 1 1 0 1 1 0 1 0 0 1   1 1 1 1 1 1 1 0 1 1 0 1 0 1 0
1 1 1 1 1 1 1 1 0 0 1 0 1 0 1   1 1 1 1 1 1 1 1 0 0 1 0 1 1 0   1 1 1 1 1 1 1 1 0 0 1 1 0 0 1
1 1 1 1 1 1 1 1 0 0 1 1 0 1 0   1 1 1 1 1 1 1 1 0 1 0 0 1 0 1   1 1 1 1 1 1 1 1 0 1 0 0 1 1 0
        1 1 1 1 1 1 1 1 0 1 0 1 0 0 1   1 1 1 1 1 1 1 1 0 1 0 1 0 1 0.
```

After removing all the overlaps, we obtain 510 distinct solutions. Notice that the coefficients d_{r0}'s ($0 \leq r \leq 7$) independently ranges over the range from 0 to 1. Therefore, we have $130560 (= 2^8 \times 510)$ distinct offsets.

6.4. 16-QAM GCSs from Non-Standard GBFs

Based on Theorem 3.3, 16-QAM GCSs from non-standard GBFs are given below.

Theorem 6.8. ([16]) For $n = m + 3$ ($m \geq 1$). Let the non-standard GBF $f(x_1, x_2, \cdots, x_m)$ come from Theorem 3.3. Again let

$$\begin{cases} A(\underline{x}) = f(x_1, x_2, \cdots, x_n) \\ a(\underline{x}) = A(\underline{x}) + s(\underline{x}) \\ B(\underline{x}) = A(\underline{x}) + \mu(\underline{x}) \\ b(\underline{x}) - a(\underline{x}) + \mu(\underline{x}), \end{cases} \tag{130}$$

$\alpha = \frac{2}{\sqrt{5}}, \beta = \frac{1}{\sqrt{5}}, j = \sqrt{-1}, \gamma = e^{j\pi/4}$, and $\xi = e^{j\pi/2}$. We construct the new 16-QAM sequences $\widetilde{A}$ and $\widetilde{B}$ with length $N = 2^n$ by

$$I: \begin{cases} \widetilde{A}(i) = \alpha\gamma j\xi^{A(i)} + \beta\gamma\xi^{a(i)} \\ \widetilde{B}(i) = \alpha\gamma\xi^{B(i)} + \beta\gamma j\xi^{b(i)} \end{cases} \text{ or } II: \begin{cases} \widetilde{A}(i) = \alpha\gamma\xi^{A(i)} + \beta\gamma\xi^{a(i)} \\ \widetilde{B}(i) = \alpha\gamma\xi^{B(i)} - \beta\gamma\xi^{b(i)}. \end{cases} \tag{131}$$

Then, the resultant 16-QAM sequences $\widetilde{A}$ and $\widetilde{B}$ form the non-standard 16-QAM GCSs if the offset $s(\underline{x})$ and the pairing difference $\mu(\underline{x})$ are given by

$$\begin{aligned} s(\underline{x}) &= d_0 + d_1 x_{\sigma(m)} \\ \mu(\underline{x}) &= 2x_{\sigma(m)}, \end{aligned} \tag{132}$$

where $2d_0 + d_1 \equiv 0 \pmod 4$ for **I** and $2d_0 + d_1 \equiv 2 \pmod 4$ for **II**.

Proof. Only derive the results in I due to similarity. For $\forall \tau > 0$, we have

$$
\begin{aligned}
& C_{\widetilde{A},\widetilde{A}}(\tau) \\
&= \sum_{i=0}^{N-\tau-1} \left[\alpha\gamma j \xi^{A(i)} + \beta\gamma \xi^{a(i)} \right] \overline{\left[\alpha\gamma j \xi^{A(l)} + \beta\gamma \xi^{a(l)} \right]} \\
&= \sum_{i=0}^{N-\tau-1} \left[\alpha^2 \xi^{A(i)-A(l)} + \beta^2 \xi^{a(i)-a(l)} + \right. \\
&\qquad\qquad\qquad \left. \alpha\beta j \left(\xi^{A(i)-a(l)} - \xi^{a(i)-A(l)} \right) \right] \\
&= \sum_{i=0}^{N-\tau-1} \left[\alpha^2 \xi^{A(i)-A(l)} + \beta^2 \xi^{a(i)-a(l)} + \right. \\
&\qquad\qquad\qquad \left. \alpha\beta j \left(\xi^{A(i)-A(l)} \xi^{-s(l)} - \xi^{A(i)-A(l)} \xi^{s(i)} \right) \right] \\
&= \sum_{i=0}^{N-\tau-1} \left[\alpha^2 \xi^{A(i)-A(l)} + \beta^2 \xi^{a(i)-a(l)} + \right. \\
&\qquad\qquad\qquad \left. \alpha\beta j \xi^{A(i)-A(l)} \left(\xi^{-s(l)} - \xi^{s(i)} \right) \right]
\end{aligned}
\tag{133}
$$

and

$$
\begin{aligned}
& C_{\widetilde{B},\widetilde{B}}(\tau) \\
&= \sum_{i=0}^{N-\tau-1} \left[\alpha\gamma \xi^{B(i)} + \beta\gamma j \xi^{b(i)} \right] \overline{\left[\alpha\gamma \xi^{B(l)} + \beta\gamma j \xi^{b(l)} \right]} \\
&= \sum_{i=0}^{N-\tau-1} \left[\alpha^2 \xi^{B(i)-B(l)} + \beta^2 \xi^{b(i)-b(l)} - \right. \\
&\qquad\qquad\qquad \left. \alpha\beta j \left(\xi^{B(i)-b(l)} - \xi^{b(i)-B(l)} \right) \right] \\
&= \sum_{i=0}^{N-\tau-1} \left[\alpha^2 \xi^{B(i)-B(l)} + \beta^2 \xi^{b(i)-b(l)} - \right. \\
&\qquad\qquad\qquad \left. \alpha\beta j \left(\xi^{B(i)-B(l)} \xi^{-s(l)} - \xi^{B(i)-B(l)} \xi^{s(i)} \right) \right] \\
&= \sum_{i=0}^{N-\tau-1} \left[\alpha^2 \xi^{B(i)-B(l)} + \beta^2 \xi^{b(i)-b(l)} - \right. \\
&\qquad\qquad\qquad \left. \alpha\beta j \xi^{B(i)-B(l)} \left(\xi^{-s(l)} - \xi^{s(i)} \right) \right] \\
&= \sum_{i=0}^{N-\tau-1} \left[\alpha^2 \xi^{B(i)-B(l)} + \beta^2 \xi^{b(i)-b(l)} - \right. \\
&\qquad\qquad \left. \alpha\beta j \xi^{A(i)-A(l)} \xi^{2i_{\sigma(m)}-2l_{\sigma(m)}} \left(\xi^{-s(l)} - \xi^{s(i)} \right) \right],
\end{aligned}
\tag{134}
$$

where $l = i + \tau$.

According to Theorem 5.3, the sequences $\underline{A}$ and $\underline{B}$ are the quaternary GCSs, and so are the sequences $\underline{a}$ and $\underline{b}$. Hence, the sum of (133) and (134) is

$$C_{\widetilde{A},\widetilde{A}}(\tau) + C_{\widetilde{B},\widetilde{B}}(\tau) = \alpha\beta j \sum_{i=0}^{N-\tau-1} \left[\xi^{A(i)-A(l)} \left(\xi^{-s(l)} \right. \right.$$
$$\left. \left. -\xi^{s(i)} \right) \left(1 - (-1)^{i_{\sigma(m)}-l_{\sigma(m)}} \right) \right]. \tag{135}$$

(1) Whenever $i_{\sigma(m)} = l_{\sigma(m)}$, we have $1 - (-1)^{i_{\sigma(m)}-l_{\sigma(m)}} = 0$, which results in that (135) vanishes.

(2) When $i_{\sigma(m)} \neq l_{\sigma(m)}$, that is, $i_{\sigma(m)} + l_{\sigma(m)} = 1$ due to the fact that $i_{\sigma(m)}$ and $l_{\sigma(m)}$ are the components of binary representations of the integers i and l, respectively, we have

$$s(l) + s(i) = 2d_0 + d_1(i_{\sigma(m)} + l_{\sigma(m)})$$
$$= 2d_0 + d_1 \equiv 0 \pmod{4}, \tag{136}$$

which implies $\xi^{-s(l)} - \xi^{s(i)} = \xi^{-s(l)}(1 - \xi^{s(l)+s(i)}) = 0$. As a consequence, (135) vanishes as well.

Summarizing the above, Theorem 6.8 is true. $\square$

Theorem 6.9. ([16]) There are $2(n-2)!(n-2)4^{n+1}$ $(n \geq 4)$ non-standard 16-QAM Golay sequences $\widetilde{A}$'s in Theorem 6.8.

Proof. Due to $2d_0 + d_1 \equiv 0 \pmod{4}$ for I, there apparently exist four solutions $(d_0, d_1) = (0,0)$, $(2,0)$, $(1,2)$ and $(3,2)$. Similarly, $(d_0, d_1) = (1,0)$, $(3,0)$, $(0,2)$ and $(2,2)$ for II. By employing Theorem 3.4, this theorem follows immediately. $\square$

Theorem 6.10. ([16]) The non-standard 16-QAM GCSs from Theorem 6.8 have the upper bound 2 of PMEPR.

Proof. Only consider I. Let the code C consist of the 16-QAM GCSs from Theorem 6.8. Thus, we have

$$|\widetilde{A}(i)|^2 = |\alpha j + \beta \xi^{s(i)}|^2 = \begin{cases} \alpha^2 + \beta^2 = 1 & s(i) = 0, 2 \\ |\alpha + \beta|^2 = 1.8 & s(i) = 1 \\ |\alpha - \beta|^2 = 0.2 & s(i) = 3, \end{cases} \tag{137}$$

and

$$|\widetilde{B}(i)|^2 = |\alpha + \beta j\xi^{s(i)}|^2 = \begin{cases} \alpha^2 + \beta^2 = 1 & s(i) = 0,2 \\ |\alpha - \beta|^2 = 0.2 & s(i) = 1 \\ |\alpha + \beta|^2 = 1.8 & s(i) = 3. \end{cases} \tag{138}$$

Therefore, the sum of AACFs of two sequences $\widetilde{\underline{A}}$ and $\widetilde{\underline{B}}$ is

$$C_{\widetilde{A},\widetilde{A}}(0) + C_{\widetilde{B},\widetilde{B}}(0) = 2(n_0 + n_1 + n_2 + n_3) = 2N, \tag{139}$$

where $n_k = |i|s(i) = k, 0 \le i \le N - 1|$ $(0 \le k \le 3)$. Hence, for $\forall \widetilde{\underline{A}} \in C$ we have $\text{PEP}(\widetilde{\underline{A}}) \le 2N$ so that $\text{PEP}(C) \le 2N$.

Assume the code C has K pairs. Due to $P_{av}(C) = \frac{2KN}{2K} = N$, $\text{PMEPR}(C) = \text{PEP}(C)/P_{av}(C) \le 2$. □

Table 4. All possible offsets and PEP upper bounds of the resultant 64-QAM GCSs in Case I

case	coefficients	w	bounds under $(abcd)$							
			(0000)	(0001)	(0010)	(0011)	(0100)	(0101)	(0110)	(0111)
			3.52	3.52	3.14	3.14	1.24	1.24	0.86	0.86
1	000000010110	$1 \le w < m$	(1000)	(1001)	(1010)	(1011)	(1100)	(1101)	(1110)	(1111)
			1.62	1.62	1.24	1.24	2.38	2.38	2	2
			(0000)	(0001)	(0010)	(0011)	(0100)	(0101)	(0110)	(0111)
			3.71	2.95	3.71	2.95	1.43	0.67	1.43	0.67
2	000000011001	$1 \le w < m$	(1000)	(1001)	(1010)	(1011)	(1100)	(1101)	(1110)	(1111)
			1.81	1.05	1.81	1.05	2.57	1.81	2.57	1.81
			(0000)	(0001)	(0010)	(0011)	(0100)	(0101)	(0110)	(0111)
			4.09	2.57	2.57	4.09	1.81	0.29	0.29	1.81
3	000000011010	$1 \le w < m$	(1000)	(1001)	(1010)	(1011)	(1100)	(1101)	(1110)	(1111)
			2.19	0.67	0.67	2.19	2.95	1.43	1.43	2.95
			(0000)	(0001)	(0010)	(0011)	(0100)	(0101)	(0110)	(0111)
			4.09	2.57	2.57	4.09	1.81	0.29	0.29	1.81
4	000000100101	$1 \le w < m$	(1000)	(1001)	(1010)	(1011)	(1100)	(1101)	(1110)	(1111)
			2.19	0.67	0.67	2.19	2.95	1.43	1.43	2.95
			(0000)	(0001)	(0010)	(0011)	(0100)	(0101)	(0110)	(0111)
			3.71	2.95	3.71	2.95	1.43	0.67	1.43	0.67
5	000000100110	$1 \le w < m$	(1000)	(1001)	(1010)	(1011)	(1100)	(1101)	(1110)	(1111)
			1.81	1.05	1.81	1.05	2.57	1.81	2.57	1.81
			(0000)	(0001)	(0010)	(0011)	(0100)	(0101)	(0110)	(0111)
			3.52	3.52	3.14	3.14	1.24	1.24	0.86	0.86
6	000000101001	$1 \le w < m$	(1000)	(1001)	(1010)	(1011)	(1100)	(1101)	(1110)	(1111)
			1.62	1.62	1.24	1.24	1.81	1.81	2	2
			(0000)	(0001)	(0010)	(0011)	(0100)	(0101)	(0110)	(0111)
			4.66	1.81	2.76	3.52	1.81	2	0.48	1.24
7	000000101010	$1 \le w \le m$	(1000)	(1001)	(1010)	(1011)	(1100)	(1101)	(1110)	(1111)
			2.76	0.48	0.86	1.62	3.52	1.24	1.62	1.81

case	coefficients	w	bounds under (abcd)							
8	000011010101	$1 \leq w < m$	(0000)	(0001)	(0010)	(0011)	(0100)	(0101)	(0110)	(0111)
			3.52	1.24	1.62	2.38	1.24	1.62	2.38	3.14
			(1000)	(1001)	(1010)	(1011)	(1100)	(1101)	(1110)	(1111)
			0.86	1.24	2	3.14	2.95	0.86	1.24	2
9	000011010110	$1 \leq w < m$	(0000)	(0001)	(0010)	(0011)	(0100)	(0101)	(0110)	(0111)
			1.81	1.81	2	2	1.81	1.81	2	2
			(1000)	(1001)	(1010)	(1011)	(1100)	(1101)	(1110)	(1111)
			2	2	1.62	1.62	2	2	1.62	1.62
10	000011011001	$1 \leq w < m$	(0000)	(0001)	(0010)	(0011)	(0100)	(0101)	(0110)	(0111)
			2.57	1.81	2.57	1.81	2.57	1.81	2.57	1.81
			(1000)	(1001)	(1010)	(1011)	(1100)	(1101)	(1110)	(1111)
			2.19	1.43	2.19	1.43	2.19	1.43	2.19	1.43
11	000011011010	$1 \leq w < m$	(0000)	(0001)	(0010)	(0011)	(0100)	(0101)	(0110)	(0111)
			2.95	1.43	1.43	2.95	2.95	1.43	1.43	2.95
			(1000)	(1001)	(1010)	(1011)	(1100)	(1101)	(1110)	(1111)
			2.57	1.05	1.05	2.57	2.57	1.05	1.05	2.57
12	000011100101	$1 \leq w < m$	(0000)	(0001)	(0010)	(0011)	(0100)	(0101)	(0110)	(0111)
			2.95	1.43	1.43	2.95	2.95	1.43	1.43	2.95
			(1000)	(1001)	(1010)	(1011)	(1100)	(1101)	(1110)	(1111)
			2.57	1.05	1.05	2.57	2.57	1.05	1.05	2.57
13	000011100110	$1 \leq w < m$	(0000)	(0001)	(0010)	(0011)	(0100)	(0101)	(0110)	(0111)
			2.57	1.81	2.57	1.81	2.57	1.81	2.57	1.81
			(1000)	(1001)	(1010)	(1011)	(1100)	(1101)	(1110)	(1111)
			2.19	1.43	2.19	1.43	2.19	1.43	2.19	1.43
14	000011101001	$1 \leq w < m$	(0000)	(0001)	(0010)	(0011)	(0100)	(0101)	(0110)	(0111)
			1.81	1.81	2	2	1.81	1.81	2	2
			(1000)	(1001)	(1010)	(1011)	(1100)	(1101)	(1110)	(1111)
			2	2	1.62	1.62	2	2	1.62	3.2
15	000011101010	$1 \leq w < m$	(0000)	(0001)	(0010)	(0011)	(0100)	(0101)	(0110)	(0111)
			3.52	1.24	1.62	1.81	3.52	1.24	1.62	1.81
			(1000)	(1001)	(1010)	(1011)	(1100)	(1101)	(1110)	(1111)
			3.14	0.86	1.24	2	3.14	0.86	1.24	2
16	001100010101	$1 \leq w < m$	(0000)	(0001)	(0010)	(0011)	(0100)	(0101)	(0110)	(0111)
			3.71	1.43	1.81	2.57	2.95	0.67	1.05	1.81
			(1000)	(1001)	(1010)	(1011)	(1100)	(1101)	(1110)	(1111)
			3.71	1.43	1.81	2.57	2.95	0.67	1.05	1.81
17	001100010110	$1 \leq w < m$	(0000)	(0001)	(0010)	(0011)	(0100)	(0101)	(0110)	(0111)
			2.57	2.57	2.19	2.19	1.81	1.81	1.43	1.43
			(1000)	(1001)	(1010)	(1011)	(1100)	(1101)	(1110)	(1111)
			2.57	2.57	2.19	2.19	1.81	1.81	1.43	1.43
18	001100011001	$1 \leq w < m$	(0000)	(0001)	(0010)	(0011)	(0100)	(0101)	(0110)	(0111)
			2.76	2	2.76	2	2	1.24	2	1.24
			(1000)	(1001)	(1010)	(1011)	(1100)	(1101)	(1110)	(1111)
			2.76	2	2.76	2	2	1.24	2	1.24
19	001100011010	$1 \leq w < m$	(0000)	(0001)	(0010)	(0011)	(0100)	(0101)	(0110)	(0111)
			3.14	1.62	1.62	3.14	1.81	0.86	0.86	1.81
			(1000)	(1001)	(1010)	(1011)	(1100)	(1101)	(1110)	(1111)
			3.14	1.62	1.62	3.14	1.81	0.86	0.86	1.81

Table 4. Continued

case	coefficients	w	bounds under ($abcd$)							
			(0000)	(0001)	(0010)	(0011)	(0100)	(0101)	(0110)	(0111)
			3.14	1.62	1.62	3.14	1.81	0.86	0.86	1.81
20	001100100101	$1 \le w < m$	(1000)	(1001)	(1010)	(1011)	(1100)	(1101)	(1110)	(1111)
			3.14	1.62	1.62	3.14	1.81	0.86	0.86	1.81
			(0000)	(0001)	(0010)	(0011)	(0100)	(0101)	(0110)	(0111)
			2.76	2	2.76	2	2	1.24	2	1.24
21	001100100110	$1 \le w < m$	(1000)	(1001)	(1010)	(1011)	(1100)	(1101)	(1110)	(1111)
			2.76	2	2.76	2	2	1.24	2	1.24
			(0000)	(0001)	(0010)	(0011)	(0100)	(0101)	(0110)	(0111)
			2.57	2.57	2.19	2.19	1.81	1.81	1.43	1.43
22	001100101001	$1 \le w < m$	(1000)	(1001)	(1010)	(1011)	(1100)	(1101)	(1110)	(1111)
			2.57	2.57	2.19	2.19	1.81	1.81	1.43	1.43
			(0000)	(0001)	(0010)	(0011)	(0100)	(0101)	(0110)	(0111)
			3.71	1.43	1.81	2.57	2.95	0.67	1.05	1.81
23	001100101010	$1 \le w < m$	(1000)	(1001)	(1010)	(1011)	(1100)	(1101)	(1110)	(1111)
			3.71	1.43	1.81	2.57	2.95	0.67	1.05	1.81
			(0000)	(0001)	(0010)	(0011)	(0100)	(0101)	(0110)	(0111)
			4.09	1.81	2.19	2.95	2.57	0.29	0.67	1.43
24	001111010101	$1 \le w < m$	(1000)	(1001)	(1010)	(1011)	(1100)	(1101)	(1110)	(1111)
			2.57	0.29	0.67	1.43	4.09	1.81	2.19	2.95
			(0000)	(0001)	(0010)	(0011)	(0100)	(0101)	(0110)	(0111)
			2.95	2.95	2.57	2.57	1.43	1.43	1.05	1.05
25	001111010110	$1 \le w < m$	(1000)	(1001)	(1010)	(1011)	(1100)	(1101)	(1110)	(1111)
			1.43	1.43	1.05	1.05	2.95	2.95	2.57	2.57
			(0000)	(0001)	(0010)	(0011)	(0100)	(0101)	(0110)	(0111)
			3.14	1.81	3.14	1.81	1.62	0.86	1.62	0.86
26	001111011001	$1 \le w < m$	(1000)	(1001)	(1010)	(1011)	(1100)	(1101)	(1110)	(1111)
			1.62	0.86	1.62	0.86	3.14	1.81	3.14	1.81
			(0000)	(0001)	(0010)	(0011)	(0100)	(0101)	(0110)	(0111)
			3.52	2	2	3.52	2	0.48	0.48	2
27	001111011010	$1 \le w < m$	(1000)	(1001)	(1010)	(1011)	(1100)	(1101)	(1110)	(1111)
			2	0.48	0.48	2	3.52	2	2	3.52
			(0000)	(0001)	(0010)	(0011)	(0100)	(0101)	(0110)	(0111)
			3.52	2	2	3.52	2	0.48	0.48	2
28	001111100101	$1 \le w < m$	(1000)	(1001)	(1010)	(1011)	(1100)	(1101)	(1110)	(1111)
			2	0.48	0.48	2	3.52	2	2	3.52
			(0000)	(0001)	(0010)	(0011)	(0100)	(0101)	(0110)	(0111)
			3.14	1.81	3.14	1.81	1.62	0.86	1.62	0.86
29	001111100110	$1 \le w < m$	(1000)	(1001)	(1010)	(1011)	(1100)	(1101)	(1110)	(1111)
			1.62	0.86	1.62	0.86	3.14	1.81	3.14	1.81
			(0000)	(0001)	(0010)	(0011)	(0100)	(0101)	(0110)	(0111)
			2.95	2.95	2.57	2.57	1.43	1.43	1.05	1.05
30	001111101001	$1 \le w < m$	(1000)	(1001)	(1010)	(1011)	(1100)	(1101)	(1110)	(1111)
			1.43	1.43	1.05	1.05	2.95	2.95	2.57	2.57
			(0000)	(0001)	(0010)	(0011)	(0100)	(0101)	(0110)	(0111)
			4.09	1.81	2.19	2.95	2.57	0.29	0.67	1.43
31	001111101010	$1 \le w < m$	(1000)	(1001)	(1010)	(1011)	(1100)	(1101)	(1110)	(1111)
			2.57	0.29	0.67	1.43	4.09	1.81	2.19	2.95

case	coefficients	w	bounds under ($abcd$)							
			(0000)	(0001)	(0010)	(0011)	(0100)	(0101)	(0110)	(0111)
32	010101000011	$1 \leq w < m$	3.52	3.52	3.14	3.14	1.24	1.24	0.86	0.86
			(1000)	(1001)	(1010)	(1011)	(1100)	(1101)	(1110)	(1111)
			1.62	1.62	1.24	1.24	1.81	1.81	2	2
			(0000)	(0001)	(0010)	(0011)	(0100)	(0101)	(0110)	(0111)
33	010101001100	$1 \leq w < m$	3.71	2.95	3.71	2.95	1.43	0.67	1.43	0.67
			(1000)	(1001)	(1010)	(1011)	(1100)	(1101)	(1110)	(1111)
			1.81	1.05	1.81	1.05	2.57	1.81	2.57	1.81
			(0000)	(0001)	(0010)	(0011)	(0100)	(0101)	(0110)	(0111)
34	010101001111	$1 \leq w < m$	4.09	2.57	2.57	4.09	1.81	0.29	0.29	1.81
			(1000)	(1001)	(1010)	(1011)	(1100)	(1101)	(1110)	(1111)
			2.19	0.67	0.67	2.19	2.95	1.43	1.43	2.95
			(0000)	(0001)	(0010)	(0011)	(0100)	(0101)	(0110)	(0111)
35	010101110000	$1 \leq w < m$	4.09	2.57	2.57	4.09	1.81	0.29	0.29	1.81
			(1000)	(1001)	(1010)	(1011)	(1100)	(1101)	(1110)	(1111)
			2.19	0.67	0.67	2.19	2.95	1.43	1.43	2.95
			(0000)	(0001)	(0010)	(0011)	(0100)	(0101)	(0110)	(0111)
36	010101110011	$1 \leq w < m$	3.71	2.95	3.71	2.95	1.43	0.67	1.43	0.67
			(1000)	(1001)	(1010)	(1011)	(1100)	(1101)	(1110)	(1111)
			1.81	1.05	1.81	1.05	2.57	1.81	2.57	1.81
			(0000)	(0001)	(0010)	(0011)	(0100)	(0101)	(0110)	(0111)
37	010101111100	$1 \leq w < m$	3.52	3.52	3.14	3.14	1.24	1.24	0.86	0.86
			(1000)	(1001)	(1010)	(1011)	(1100)	(1101)	(1110)	(1111)
			1.62	1.62	1.24	1.24	1.81	1.81	2	2
			(0000)	(0001)	(0010)	(0011)	(0100)	(0101)	(0110)	(0111)
38	01010111111	$1 \leq w < m$	4.66	1.81	2.76	3.52	1.81	2	0.48	1.24
			(1000)	(1001)	(1010)	(1011)	(1100)	(1101)	(1110)	(1111)
			2.76	0.48	0.86	1.62	3.52	1.24	1.62	1.81
			(0000)	(0001)	(0010)	(0011)	(0100)	(0101)	(0110)	(0111)
39	010110000000	$1 \leq w < m$	3.52	1.24	1.62	1.81	3.52	1.24	1.62	1.81
			(1000)	(1001)	(1010)	(1011)	(1100)	(1101)	(1110)	(1111)
			3.14	0.86	1.24	2	3.14	0.86	1.24	2
			(0000)	(0001)	(0010)	(0011)	(0100)	(0101)	(0110)	(0111)
40	010110000011	$1 \leq w < m$	1.81	1.81	2	2	1.81	1.81	2	2
			(1000)	(1001)	(1010)	(1011)	(1100)	(1101)	(1110)	(1111)
			2	2	1.62	1.62	2	2	1.62	1.62
			(0000)	(0001)	(0010)	(0011)	(0100)	(0101)	(0110)	(0111)
41	010110001100	$1 \leq w < m$	2.57	1.81	2.57	1.81	2.57	1.81	2.57	1.81
			(1000)	(1001)	(1010)	(1011)	(1100)	(1101)	(1110)	(1111)
			2.19	1.43	2.19	1.43	2.19	1.43	2.19	1.43
			(0000)	(0001)	(0010)	(0011)	(0100)	(0101)	(0110)	(0111)
42	010110001111	$1 \leq w < m$	2.95	1.43	1.43	2.95	2.95	1.43	1.43	2.95
			(1000)	(1001)	(1010)	(1011)	(1100)	(1101)	(1110)	(1111)
			2.57	1.05	1.05	2.57	2.57	1.05	1.05	2.57
			(0000)	(0001)	(0010)	(0011)	(0100)	(0101)	(0110)	(0111)
43	010110110000	$1 \leq w < m$	.95	1.43	1.43	2.95	2.95	1.43	1.43	2.95
			(1000)	(1001)	(1010)	(1011)	(1100)	(1101)	(1110)	(1111)
			2.57	1.05	1.05	2.57	2.57	1.05	1.05	2.57

Table 4. Continued

case	coefficients	w	bounds under (abcd)							
			(0000)	(0001)	(0010)	(0011)	(0100)	(0101)	(0110)	(0111)
44	010110110011	$1 \le w < m$	2.57	1.81	2.57	1.81	2.57	1.81	2.57	1.81
			(1000)	(1001)	(1010)	(1011)	(1100)	(1101)	(1110)	(1111)
			2.19	1.43	2.19	1.43	2.19	1.43	2.19	1.43
			(0000)	(0001)	(0010)	(0011)	(0100)	(0101)	(0110)	(0111)
45	010110111100	$1 \le w < m$	1.81	1.81	2	2	1.81	1.81	2	2
			(1000)	(1001)	(1010)	(1011)	(1100)	(1101)	(1110)	(1111)
			2	2	1.62	1.62	2	2	1.62	1.62
			(0000)	(0001)	(0010)	(0011)	(0100)	(0101)	(0110)	(0111)
46	010110111111	$1 \le w < m$	3.52	1.24	1.62	1.81	3.52	1.24	1.62	1.81
			(1000)	(1001)	(1010)	(1011)	(1100)	(1101)	(1110)	(1111)
			3.14	0.86	1.24	2	3.14	0.86	1.24	2
			(0000)	(0001)	(0010)	(0011)	(0100)	(0101)	(0110)	(0111)
47	011001000000	$1 \le w < m$	3.71	1.43	1.81	2.57	2.95	0.67	1.05	1.81
			(1000)	(1001)	(1010)	(1011)	(1100)	(1101)	(1110)	(1111)
			3.71	1.43	1.81	2.57	2.95	0.67	1.05	1.81
			(0000)	(0001)	(0010)	(0011)	(0100)	(0101)	(0110)	(0111)
48	011001000011	$1 \le w < m$	2.57	2.57	2.19	2.19	1.81	1.81	1.43	1.43
			(1000)	(1001)	(1010)	(1011)	(1100)	(1101)	(1110)	(1111)
			2.57	2.57	2.19	2.19	1.81	1.81	1.43	1.43
			(0000)	(0001)	(0010)	(0011)	(0100)	(0101)	(0110)	(0111)
49	011001001100	$1 \le w < m$	2.76	2	2.76	2	2	1.24	2	1.24
			(1000)	(1001)	(1010)	(1011)	(1100)	(1101)	(1110)	(1111)
			2.76	2	2.76	2	2	1.24	2	1.24
			(0000)	(0001)	(0010)	(0011)	(0100)	(0101)	(0110)	(0111)
50	011001001111	$1 \le w < m$	3.14	1.62	1.62	3.14	1.81	0.86	0.86	1.81
			(1000)	(1001)	(1010)	(1011)	(1100)	(1101)	(1110)	(1111)
			3.14	1.62	1.62	3.14	1.81	0.86	0.86	1.81
			(0000)	(0001)	(0010)	(0011)	(0100)	(0101)	(0110)	(0111)
51	011001110000	$1 \le w < m$	3.14	1.62	1.62	3.14	1.81	0.86	0.86	1.81
			(1000)	(1001)	(1010)	(1011)	(1100)	(1101)	(1110)	(1111)
			3.14	1.62	1.62	3.14	1.81	0.86	0.86	1.81
			(0000)	(0001)	(0010)	(0011)	(0100)	(0101)	(0110)	(0111)
52	011001110011	$1 \le w < m$	2.76	2	2.76	2	2	1.24	2	1.24
			(1000)	(1001)	(1010)	(1011)	(1100)	(1101)	(1110)	(1111)
			2.76	2	2.76	2	2	1.24	2	1.24
			(0000)	(0001)	(0010)	(0011)	(0100)	(0101)	(0110)	(0111)
53	011001111100	$1 \le w < m$	2.57	2.57	2.19	2.19	1.81	1.81	1.43	1.43
			(1000)	(1001)	(1010)	(1011)	(1100)	(1101)	(1110)	(1111)
			2.57	2.57	2.19	2.19	1.81	1.81	1.43	1.43
			(0000)	(0001)	(0010)	(0011)	(0100)	(0101)	(0110)	(0111)
54	011001111111	$1 \le w < m$	3.71	1.43	1.81	2.57	2.95	0.67	1.05	1.81
			(1000)	(1001)	(1010)	(1011)	(1100)	(1101)	(1110)	(1111)
			3.71	1.43	1.81	2.57	2.95	0.67	1.05	1.81
			(0000)	(0001)	(0010)	(0011)	(0100)	(0101)	(0110)	(0111)
55	011010000000	$1 \le w < m$	4.09	1.81	2.19	2.95	2.57	0.29	0.67	1.43
			(1000)	(1001)	(1010)	(1011)	(1100)	(1101)	(1110)	(1111)
			2.57	0.29	0.67	1.43	4.09	1.81	2.19	2.95

case	coefficients	w	bounds under ($abcd$)							
56	011010000011	$1 \leq w < m$	(0000)	(0001)	(0010)	(0011)	(0100)	(0101)	(0110)	(0111)
			2.95	2.95	2.57	2.57	1.43	1.43	1.05	1.05
			(1000)	(1001)	(1010)	(1011)	(1100)	(1101)	(1110)	(1111)
			1.43	1.43	1.05	1.05	2.95	2.95	2.57	2.57
57	011010001100	$1 \leq w < m$	(0000)	(0001)	(0010)	(0011)	(0100)	(0101)	(0110)	(0111)
			3.14	1.81	3.14	1.81	1.62	0.86	1.62	0.86
			(1000)	(1001)	(1010)	(1011)	(1100)	(1101)	(1110)	(1111)
			1.62	0.86	1.62	0.86	3.14	1.81	3.14	1.81
58	011010001111	$1 \leq w < m$	(0000)	(0001)	(0010)	(0011)	(0100)	(0101)	(0110)	(0111)
			3.52	2	2	3.52	2	0.48	0.48	2
			(1000)	(1001)	(1010)	(1011)	(1100)	(1101)	(1110)	(1111)
			2	0.48	0.48	2	3.52	2	2	3.52
59	011010110000	$1 \leq w < m$	(0000)	(0001)	(0010)	(0011)	(0100)	(0101)	(0110)	(0111)
			3.52	2	2	3.52	2	0.48	0.48	2
			(1000)	(1001)	(1010)	(1011)	(1100)	(1101)	(1110)	(1111)
			2	0.48	0.48	2	3.52	2	2	3.52
60	011010110011	$1 \leq w < m$	(0000)	(0001)	(0010)	(0011)	(0100)	(0101)	(0110)	(0111)
			3.14	1.81	3.14	1.81	1.62	0.86	1.62	0.86
			(1000)	(1001)	(1010)	(1011)	(1100)	(1101)	(1110)	(1111)
			1.62	0.86	1.62	0.86	3.14	1.81	3.14	1.81
61	011010111100	$1 \leq w < m$	(0000)	(0001)	(0010)	(0011)	(0100)	(0101)	(0110)	(0111)
			2.95	2.95	2.57	2.57	1.43	1.43	1.05	1.05
			(1000)	(1001)	(1010)	(1011)	(1100)	(1101)	(1110)	(1111)
			1.43	1.43	1.05	1.05	2.95	2.95	2.57	2.57
62	011010111111	$1 \leq w < m$	(0000)	(0001)	(0010)	(0011)	(0100)	(0101)	(0110)	(0111)
			4.09	1.81	2.19	2.95	2.57	0.29	0.67	1.43
			(1000)	(1001)	(1010)	(1011)	(1100)	(1101)	(1110)	(1111)
			2.57	0.29	0.67	1.43	4.09	1.81	2.19	2.95
63	100101000000	$1 \leq w < m$	(0000)	(0001)	(0010)	(0011)	(0100)	(0101)	(0110)	(0111)
			4.09	1.81	2.19	2.95	2.57	0.29	0.67	1.43
			(1000)	(1001)	(1010)	(1011)	(1100)	(1101)	(1110)	(1111)
			2.57	0.29	0.67	1.43	4.09	1.81	2.19	2.95
64	100101000011	$1 \leq w < m$	(0000)	(0001)	(0010)	(0011)	(0100)	(0101)	(0110)	(0111)
			2.95	2.95	2.57	2.57	1.43	1.43	1.05	1.05
			(1000)	(1001)	(1010)	(1011)	(1100)	(1101)	(1110)	(1111)
			1.43	1.43	1.05	1.05	2.95	2.95	2.57	2.57
65	100101001100	$1 \leq w < m$	(0000)	(0001)	(0010)	(0011)	(0100)	(0101)	(0110)	(0111)
			3.14	1.81	3.14	1.81	1.62	0.86	1.62	0.86
			(1000)	(1001)	(1010)	(1011)	(1100)	(1101)	(1110)	(1111)
			1.62	0.86	1.62	0.86	3.14	1.81	3.14	1.81
66	100101001111	$1 \leq w < m$	(0000)	(0001)	(0010)	(0011)	(0100)	(0101)	(0110)	(0111)
			3.52	2	2	3.52	2	0.48	0.48	2
			(1000)	(1001)	(1010)	(1011)	(1100)	(1101)	(1110)	(1111)
			2	0.48	0.48	2	3.52	4	2	3.52
67	100101110000	$1 \leq w < m$	(0000)	(0001)	(0010)	(0011)	(0100)	(0101)	(0110)	(0111)
			3.52	2	2	3.52	2	0.48	0.48	2
			(1000)	(1001)	(1010)	(1011)	(1100)	(1101)	(1110)	(1111)
			2	0.48	0.48	2	3.52	2	2	3.52

Table 4. Continued

case	coefficients	w	bounds under ($abcd$)							
			(0000)	(0001)	(0010)	(0011)	(0100)	(0101)	(0110)	(0111)
68	100101110011	$1 \le w < m$	3.14	1.81	3.14	1.81	1.62	0.86	1.62	0.86
			(1000)	(1001)	(1010)	(1011)	(1100)	(1101)	(1110)	(1111)
			1.62	0.86	1.62	0.86	3.14	1.81	3.14	1.81
			(0000)	(0001)	(0010)	(0011)	(0100)	(0101)	(0110)	(0111)
69	100101111100	$1 \le w < m$	2.95	2.95	2.57	2.57	1.43	1.43	1.05	1.05
			(1000)	(1001)	(1010)	(1011)	(1100)	(1101)	(1110)	(1111)
			1.43	1.43	1.05	1.05	2.95	2.95	2.57	2.57
			(0000)	(0001)	(0010)	(0011)	(0100)	(0101)	(0110)	(0111)
70	10010111111	$1 \le w < m$	4.09	1.81	2.19	2.95	2.57	0.29	0.67	1.43
			(1000)	(1001)	(1010)	(1011)	(1100)	(1101)	(1110)	(1111)
			2.57	0.29	0.67	1.43	4.09	1.81	2.19	2.95
			(0000)	(0001)	(0010)	(0011)	(0100)	(0101)	(0110)	(0111)
71	100110000000	$1 \le w < m$	3.71	1.43	1.81	2.57	2.95	0.67	1.05	1.81
			(1000)	(1001)	(1010)	(1011)	(1100)	(1101)	(1110)	(1111)
			3.71	1.43	1.81	2.57	2.95	0.67	1.05	1.81
			(0000)	(0001)	(0010)	(0011)	(0100)	(0101)	(0110)	(0111)
72	100110000011	$1 \le w < m$	2.57	2.57	2.19	2.19	1.81	1.81	1.43	1.43
			(1000)	(1001)	(1010)	(1011)	(1100)	(1101)	(1110)	(1111)
			2.57	2.57	2.19	2.19	1.81	1.81	1.43	1.43
			(0000)	(0001)	(0010)	(0011)	(0100)	(0101)	(0110)	(0111)
73	100110001100	$1 \le w < m$	2.76	2	2.76	2	2	1.24	2	1.24
			(1000)	(1001)	(1010)	(1011)	(1100)	(1101)	(1110)	(1111)
			2.76	2	2.76	2	2	1.24	2	1.24
			(0000)	(0001)	(0010)	(0011)	(0100)	(0101)	(0110)	(0111)
74	100110001111	$1 \le w < m$	3.14	1.62	1.62	3.14	1.81	0.86	0.86	1.81
			(1000)	(1001)	(1010)	(1011)	(1100)	(1101)	(1110)	(1111)
			3.14	1.62	1.62	3.14	1.81	0.86	0.86	1.81
			(0000)	(0001)	(0010)	(0011)	(0100)	(0101)	(0110)	(0111)
75	100110110000	$1 \le w < m$	3.14	1.62	1.62	3.14	1.81	0.86	0.86	1.81
			(1000)	(1001)	(1010)	(1011)	(1100)	(1101)	(1110)	(1111)
			3.14	1.62	1.62	3.14	1.81	0.86	0.86	1.81
			(0000)	(0001)	(0010)	(0011)	(0100)	(0101)	(0110)	(0111)
76	100110110011	$1 \le w < m$	2.76	2	2.76	2	2	1.24	2	1.24
			(1000)	(1001)	(1010)	(1011)	(1100)	(1101)	(1110)	(1111)
			2.76	2	2.76	2	2	1.24	2	1.24
			(0000)	(0001)	(0010)	(0011)	(0100)	(0101)	(0110)	(0111)
77	100110111100	$1 \le w < m$	2.57	2.57	2.19	2.19	1.81	1.81	1.43	1.43
			(1000)	(1001)	(1010)	(1011)	(1100)	(1101)	(1110)	(1111)
			2.57	2.57	2.19	2.19	1.81	1.81	1.43	1.43
			(0000)	(0001)	(0010)	(0011)	(0100)	(0101)	(0110)	(0111)
78	10011011111	$1 \le w < m$	3.71	1.43	1.81	2.57	2.95	0.67	1.05	1.81
			(1000)	(1001)	(1010)	(1011)	(1100)	(1101)	(1110)	(1111)
			3.71	1.43	1.81	2.57	2.95	0.67	1.05	1.81
			(0000)	(0001)	(0010)	(0011)	(0100)	(0101)	(0110)	(0111)
79	101001000000	$1 \le w < m$	3.52	1.24	1.62	1.81	3.52	1.24	1.62	1.81
			(1000)	(1001)	(1010)	(1011)	(1100)	(1101)	(1110)	(1111)
			3.14	0.86	1.24	2	3.14	0.86	1.24	2

case	coefficients	w	bounds under ($abcd$)							
			(0000)	(0001)	(0010)	(0011)	(0100)	(0101)	(0110)	(0111)
80	101001000011	$1 \leq w < m$	1.81	1.81	2	2	1.81	1.81	2	2
			(1000)	(1001)	(1010)	(1011)	(1100)	(1101)	(1110)	(1111)
			2	2	1.62	1.62	2	2	1.62	1.62
			(0000)	(0001)	(0010)	(0011)	(0100)	(0101)	(0110)	(0111)
81	101001001100	$1 \leq w < m$	2.57	1.81	2.57	1.81	2.57	1.81	2.57	1.81
			(1000)	(1001)	(1010)	(1011)	(1100)	(1101)	(1110)	(1111)
			2.19	1.43	2.19	1.43	2.19	1.43	2.19	1.43
			(0000)	(0001)	(0010)	(0011)	(0100)	(0101)	(0110)	(0111)
82	101001001111	$1 \leq w < m$	2.95	1.43	1.43	2.95	2.95	1.43	1.43	2.95
			(1000)	(1001)	(1010)	(1011)	(1100)	(1101)	(1110)	(1111)
			2.57	1.05	1.05	2.57	2.57	1.05	1.05	2.57
			(0000)	(0001)	(0010)	(0011)	(0100)	(0101)	(0110)	(0111)
83	101001110000	$1 \leq w < m$	2.95	1.43	1.43	2.95	2.95	1.43	1.43	2.95
			(1000)	(1001)	(1010)	(1011)	(1100)	(1101)	(1110)	(1111)
			2.57	1.05	1.05	2.57	2.57	1.05	1.05	2.57
			(0000)	(0001)	(0010)	(0011)	(0100)	(0101)	(0110)	(0111)
84	101001110011	$1 \leq w < m$	2.57	1.81	2.57	1.81	2.57	1.81	2.57	1.81
			(1000)	(1001)	(1010)	(1011)	(1100)	(1101)	(1110)	(1111)
			2.19	1.43	2.19	1.43	2.19	1.43	2.19	1.43
			(0000)	(0001)	(0010)	(0011)	(0100)	(0101)	(0110)	(0111)
85	101001111100	$1 \leq w < m$	1.81	1.81	2	2	1.81	1.81	2	2
			(1000)	(1001)	(1010)	(1011)	(1100)	(1101)	(1110)	(1111)
			2	2	1.62	1.62	2	2	1.62	1.62
			(0000)	(0001)	(0010)	(0011)	(0100)	(0101)	(0110)	(0111)
86	101001111111	$1 \leq w < m$	3.52	1.24	1.62	1.81	3.52	1.24	1.62	1.81
			(1000)	(1001)	(1010)	(1011)	(1100)	(1101)	(1110)	(1111)
			3.14	0.86	1.24	2	3.14	0.86	1.24	2
			(0000)	(0001)	(0010)	(0011)	(0100)	(0101)	(0110)	(0111)
87	101010000000	$1 \leq w \leq m$	4.66	1.81	2.76	3.52	1.81	2	0.48	1.24
			(1000)	(1001)	(1010)	(1011)	(1100)	(1101)	(1110)	(1111)
			2.76	0.48	0.86	1.62	3.52	1.24	1.62	1.81
			(0000)	(0001)	(0010)	(0011)	(0100)	(0101)	(0110)	(0111)
88	101010000011	$1 \leq w < m$	3.52	3.52	3.14	3.14	1.24	1.24	0.86	0.86
			(1000)	(1001)	(1010)	(1011)	(1100)	(1101)	(1110)	(1111)
			1.62	1.62	1.24	1.24	1.81	1.81	2	2
			(0000)	(0001)	(0010)	(0011)	(0100)	(0101)	(0110)	(0111)
89	101010001100	$1 \leq w < m$	3.71	2.95	3.71	2.95	1.43	0.67	1.43	0.67
			(1000)	(1001)	(1010)	(1011)	(1100)	(1101)	(1110)	(1111)
			1.81	1.05	1.81	1.05	2.57	1.81	2.57	1.81
			(0000)	(0001)	(0010)	(0011)	(0100)	(0101)	(0110)	(0111)
90	101010001111	$1 \leq w < m$	4.09	2.57	2.57	4.09	1.81	0.29	0.29	1.81
			(1000)	(1001)	(1010)	(1011)	(1100)	(1101)	(1110)	(1111)
			2.19	0.67	0.67	2.19	2.95	1.43	1.43	2.95
			(0000)	(0001)	(0010)	(0011)	(0100)	(0101)	(0110)	(0111)
91	101010110000	$1 \leq w < m$	4.09	2.57	2.57	4.09	1.81	0.29	0.29	1.81
			(1000)	(1001)	(1010)	(1011)	(1100)	(1101)	(1110)	(1111)
			2.19	0.67	0.67	2.19	2.95	1.43	1.43	2.95

Table 4. Continued

case	coefficients	w	bounds under ($abcd$)							
			(0000)	(0001)	(0010)	(0011)	(0100)	(0101)	(0110)	(0111)
92	101010110011	$1 \leq w < m$	3.71	2.95	3.71	2.95	1.43	0.67	1.43	0.67
			(1000)	(1001)	(1010)	(1011)	(1100)	(1101)	(1110)	(1111)
			1.81	1.05	1.81	1.05	2.57	1.81	2.57	1.81
			(0000)	(0001)	(0010)	(0011)	(0100)	(0101)	(0110)	(0111)
93	101010111100	$1 \leq w < m$	3.52	3.52	3.14	3.14	1.24	1.24	0.86	0.86
			(1000)	(1001)	(1010)	(1011)	(1100)	(1101)	(1110)	(1111)
			1.62	1.62	1.24	1.24	1.81	1.81	2	2
			(0000)	(0001)	(0010)	(0011)	(0100)	(0101)	(0110)	(0111)
94	101010111111	$1 \leq w < m$	4.66	1.81	2.76	3.52	1.81	2	0.48	1.24
			(1000)	(1001)	(1010)	(1011)	(1100)	(1101)	(1110)	(1111)
			2.76	0.48	0.86	1.62	3.52	1.24	1.62	1.81
			(0000)	(0001)	(0010)	(0011)	(0100)	(0101)	(0110)	(0111)
95	110000010101	$1 \leq w < m$	4.09	1.81	2.19	2.95	2.57	0.29	0.67	1.43
			(1000)	(1001)	(1010)	(1011)	(1100)	(1101)	(1110)	(1111)
			2.57	0.29	0.67	1.43	4.09	1.81	2.19	2.95
			(0000)	(0001)	(0010)	(0011)	(0100)	(0101)	(0110)	(0111)
96	110000010110	$1 \leq w < m$	2.95	2.95	2.57	2.57	1.43	1.43	1.05	1.05
			(1000)	(1001)	(1010)	(1011)	(1100)	(1101)	(1110)	(1111)
			1.43	1.43	1.05	1.05	2.95	2.95	2.57	2.57
			(0000)	(0001)	(0010)	(0011)	(0100)	(0101)	(0110)	(0111)
97	110000011001	$1 \leq w < m$	3.14	1.81	3.14	1.81	1.62	0.86	1.62	0.86
			(1000)	(1001)	(1010)	(1011)	(1100)	(1101)	(1110)	(1111)
			1.62	0.86	1.62	0.86	3.14	1.81	3.14	1.81
			(0000)	(0001)	(0010)	(0011)	(0100)	(0101)	(0110)	(0111)
98	110000011010	$1 \leq w < m$	3.52	2	2	3.52	2	0.48	0.48	2
			(1000)	(1001)	(1010)	(1011)	(1100)	(1101)	(1110)	(1111)
			2	0.48	0.48	2	3.52	2	2	3.52
			(0000)	(0001)	(0010)	(0011)	(0100)	(0101)	(0110)	(0111)
99	110000100101	$1 \leq w < m$	3.52	2	2	3.52	2	0.48	0.48	2
			(1000)	(1001)	(1010)	(1011)	(1100)	(1101)	(1110)	(1111)
			2	0.48	0.48	2	3.52	2	2	3.52
			(0000)	(0001)	(0010)	(0011)	(0100)	(0101)	(0110)	(0111)
100	110000100110	$1 \leq w < m$	3.14	1.81	3.14	1.81	1.62	0.86	1.62	0.86
			(1000)	(1001)	(1010)	(1011)	(1100)	(1101)	(1110)	(1111)
			1.62	0.86	1.62	0.86	3.14	1.81	3.14	1.81
			(0000)	(0001)	(0010)	(0011)	(0100)	(0101)	(0110)	(0111)
101	110000101001	$1 \leq w < m$	2.95	2.95	2.57	2.57	1.43	1.43	1.05	1.05
			(1000)	(1001)	(1010)	(1011)	(1100)	(1101)	(1110)	(1111)
			1.43	1.43	1.05	1.05	2.95	2.95	2.57	2.57
			(0000)	(0001)	(0010)	(0011)	(0100)	(0101)	(0110)	(0111)
102	110000101010	$1 \leq w < m$	4.09	1.81	2.19	2.95	2.57	0.29	0.67	1.43
			(1000)	(1001)	(1010)	(1011)	(1100)	(1101)	(1110)	(1111)
			2.57	0.29	0.67	1.43	4.09	1.81	2.19	2.95
			(0000)	(0001)	(0010)	(0011)	(0100)	(0101)	(0110)	(0111)
103	110011010101	$1 \leq w < m$	3.71	1.43	1.81	2.57	2.95	0.67	1.05	1.81
			(1000)	(1001)	(1010)	(1011)	(1100)	(1101)	(1110)	(1111)
			3.71	1.43	1.81	2.57	2.95	0.67	1.05	1.81

case	coefficients	w	bounds under ($abcd$)							
			(0000)	(0001)	(0010)	(0011)	(0100)	(0101)	(0110)	(0111)
104	110011010110	$1 \leq w < m$	2.57	2.57	2.19	2.19	1.81	1.81	1.43	1.43
			(1000)	(1001)	(1010)	(1011)	(1100)	(1101)	(1110)	(1111)
			2.57	2.57	2.19	2.19	1.81	1.81	1.43	1.43
			(0000)	(0001)	(0010)	(0011)	(0100)	(0101)	(0110)	(0111)
105	110011011001	$1 \leq w < m$	2.76	2	2.76	2	2	1.24	2	1.24
			(1000)	(1001)	(1010)	(1011)	(1100)	(1101)	(1110)	(1111)
			2.76	2	2.76	2	2	1.24	2	1.24
			(0000)	(0001)	(0010)	(0011)	(0100)	(0101)	(0110)	(0111)
106	110011011010	$1 \leq w < m$	3.14	1.62	1.62	3.14	1.81	0.86	0.86	1.81
			(1000)	(1001)	(1010)	(1011)	(1100)	(1101)	(1110)	(1111)
			3.14	1.62	1.62	3.14	1.81	0.86	0.86	1.81
			(0000)	(0001)	(0010)	(0011)	(0100)	(0101)	(0110)	(0111)
107	110011100101	$1 \leq w < m$	3.14	1.62	1.62	3.14	1.81	0.86	0.86	1.81
			(1000)	(1001)	(1010)	(1011)	(1100)	(1101)	(1110)	(1111)
			3.14	1.62	1.62	3.14	1.81	0.86	0.86	1.81
			(0000)	(0001)	(0010)	(0011)	(0100)	(0101)	(0110)	(0111)
108	110011100110	$1 \leq w < m$	2.76	2	2.76	2	2	1.24	2	1.24
			(1000)	(1001)	(1010)	(1011)	(1100)	(1101)	(1110)	(1111)
			2.76	2	2.76	2	2	1.24	2	1.24
			(0000)	(0001)	(0010)	(0011)	(0100)	(0101)	(0110)	(0111)
109	110011101001	$1 \leq w < m$	2.57	2.57	2.19	2.19	1.81	1.81	1.43	1.43
			(1000)	(1001)	(1010)	(1011)	(1100)	(1101)	(1110)	(1111)
			3.14	2.57	2.19	2.19	1.81	1.81	1.43	1.43
			(0000)	(0001)	(0010)	(0011)	(0100)	(0101)	(0110)	(0111)
110	110011101010	$1 \leq w < m$	3.71	1.43	1.81	2.57	2.95	0.67	1.05	1.81
			(1000)	(1001)	(1010)	(1011)	(1100)	(1101)	(1110)	(1111)
			3.71	1.43	1.81	2.57	2.95	0.67	1.05	1.81
			(0000)	(0001)	(0010)	(0011)	(0100)	(0101)	(0110)	(0111)
111	111100010101	$1 \leq w < m$	3.52	1.24	1.62	1.81	3.52	1.24	1.62	1.81
			(1000)	(1001)	(1010)	(1011)	(1100)	(1101)	(1110)	(1111)
			3.14	0.86	1.24	2	3.14	0.86	1.24	2
			(0000)	(0001)	(0010)	(0011)	(0100)	(0101)	(0110)	(0111)
112	111100010110	$1 \leq w < m$	1.81	1.81	2	2	1.81	1.81	2	2
			(1000)	(1001)	(1010)	(1011)	(1100)	(1101)	(1110)	(1111)
			2	2	1.62	1.62	2	2	1.62	1.62
			(0000)	(0001)	(0010)	(0011)	(0100)	(0101)	(0110)	(0111)
113	111100011001	$1 \leq w < m$	2.57	1.81	2.57	1.81	2.57	1.81	2.57	1.81
			(1000)	(1001)	(1010)	(1011)	(1100)	(1101)	(1110)	(1111)
			2.19	1.43	2.19	1.43	2.19	1.43	2.19	1.43
			(0000)	(0001)	(0010)	(0011)	(0100)	(0101)	(0110)	(0111)
114	111100011010	$1 \leq w < m$	2.95	1.43	1.43	2.95	2.95	1.43	1.43	2.95
			(1000)	(1001)	(1010)	(1011)	(1100)	(1101)	(1110)	(1111)
			2.57	1.05	1.05	2.57	2.57	1.05	1.05	2.57
			(0000)	(0001)	(0010)	(0011)	(0100)	(0101)	(0110)	(0111)
115	111100100101	$1 \leq w < m$	2.95	1.43	1.43	2.95	2.95	1.43	1.43	2.95
			(1000)	(1001)	(1010)	(1011)	(1100)	(1101)	(1110)	(1111)
			2.57	1.05	1.05	2.57	2.57	1.05	1.05	2.57

Table 4. Continued

case	coefficients	w	bounds under ($abcd$)							
			(0000)	(0001)	(0010)	(0011)	(0100)	(0101)	(0110)	(0111)
116	111100100110	$1 \leq w < m$	2.57	1.81	2.57	1.81	2.57	1.81	2.57	1.81
			(1000)	(1001)	(1010)	(1011)	(1100)	(1101)	(1110)	(1111)
			2.19	1.43	2.19	1.43	2.19	1.43	2.19	1.43
117	111100101001	$1 \leq w < m$	(0000)	(0001)	(0010)	(0011)	(0100)	(0101)	(0110)	(0111)
			1.81	1.81	2	2	1.81	1.81	2	2
			(1000)	(1001)	(1010)	(1011)	(1100)	(1101)	(1110)	(1111)
			2	2	1.62	1.62	2	2	1.62	1.62
118	111100101010	$1 \leq w < m$	(0000)	(0001)	(0010)	(0011)	(0100)	(0101)	(0110)	(0111)
			3.52	1.24	1.62	1.81	3.52	1.24	1.62	1.81
			(1000)	(1001)	(1010)	(1011)	(1100)	(1101)	(1110)	(1111)
			3.14	0.86	1.24	2	3.14	0.86	1.24	2
119	111111010101	$1 \leq w < m$	(0000)	(0001)	(0010)	(0011)	(0100)	(0101)	(0110)	(0111)
			4.66	1.81	2.76	3.52	1.81	2	0.48	1.24
			(1000)	(1001)	(1010)	(1011)	(1100)	(1101)	(1110)	(1111)
			2.76	0.48	0.86	1.62	3.52	1.24	1.62	1.81
120	111111010110	$1 \leq w < m$	(0000)	(0001)	(0010)	(0011)	(0100)	(0101)	(0110)	(0111)
			3.52	3.52	3.14	3.14	1.24	1.24	0.86	0.86
			(1000)	(1001)	(1010)	(1011)	(1100)	(1101)	(1110)	(1111)
			1.62	1.62	1.24	1.24	1.81	1.81	2	2
121	111111011001	$1 \leq w < m$	(0000)	(0001)	(0010)	(0011)	(0100)	(0101)	(0110)	(0111)
			3.71	2.95	3.71	2.95	1.43	0.67	1.43	0.67
			(1000)	(1001)	(1010)	(1011)	(1100)	(1101)	(1110)	(1111)
			1.81	1.05	1.81	1.05	2.57	1.81	2.57	1.81
122	111111011010	$1 \leq w < m$	(0000)	(0001)	(0010)	(0011)	(0100)	(0101)	(0110)	(0111)
			2.57	2.57	4.09	1.81	0.29	0.29	1.81	2.19
			(1000)	(1001)	(1010)	(1011)	(1100)	(1101)	(1110)	(1111)
			2.19	0.67	0.67	2.19	2.95	1.43	1.43	2.95
123	111111100101	$1 \leq w < m$	(0000)	(0001)	(0010)	(0011)	(0100)	(0101)	(0110)	(0111)
			4.09	2.57	2.57	4.09	1.81	0.29	0.29	1.81
			(1000)	(1001)	(1010)	(1011)	(1100)	(1101)	(1110)	(1111)
			2.19	0.67	0.67	2.19	2.95	1.43	1.43	2.95
124	111111100110	$1 \leq w < m$	(0000)	(0001)	(0010)	(0011)	(0100)	(0101)	(0110)	(0111)
			3.71	2.95	3.71	2.95	1.43	0.67	1.43	0.67
			(1000)	(1001)	(1010)	(1011)	(1100)	(1101)	(1110)	(1111)
			1.81	1.05	1.81	1.05	2.57	1.81	2.57	1.81
125	111111101001	$1 \leq w < m$	(0000)	(0001)	(0010)	(0011)	(0100)	(0101)	(0110)	(0111)
			3.52	3.52	3.14	3.14	1.24	1.24	0.86	0.86
			(1000)	(1001)	(1010)	(1011)	(1100)	(1101)	(1110)	(1111)
			1.62	1.62	1.24	1.24	1.81	1.81	2	2
126	111111101010	$1 \leq w < m$	(0000)	(0001)	(0010)	(0011)	(0100)	(0101)	(0110)	(0111)
			4.66	1.81	2.76	3.52	1.81	2	0.48	1.24
			(1000)	(1001)	(1010)	(1011)	(1100)	(1101)	(1110)	(1111)
			2.76	0.48	0.86	1.62	3.52	1.24	1.62	1.81

Table 5. All possible offsets and PEP upper bounds of the resultant 64-QAM GCSs in Cases II and III

case	coefficients	w	bounds under $(abcd)$							
1	000000	$1/m$	(0000)	(0001)	(0010)	(0011)	(0100)	(0101)	(0110)	(0111)
			4.66	2.38	2.76	3.52	2.38	0.19	0.48	1.24
			(1000)	(1001)	(1010)	(1011)	(1100)	(1101)	(1110)	(1111)
			2.76	0.48	0.86	1.62	3.52	1.24	1.62	2.38
2	000001	$1/m$	(0000)	(0001)	(0010)	(0011)	(0100)	(0101)	(0110)	(0111)
			3.52	3.52	3.14	3.14	1.24	1.24	0.86	0.86
			(1000)	(1001)	(1010)	(1011)	(1100)	(1101)	(1110)	(1111)
			1.62	1.62	1.24	1.24	2.38	2.38	2	2
3	000010	$1/m$	(0000)	(0001)	(0010)	(0011)	(0100)	(0101)	(0110)	(0111)
			3.71	2.95	3.71	2.95	1.43	0.67	1.43	0.67
			(1000)	(1001)	(1010)	(1011)	(1100)	(1101)	(1110)	(1111)
			1.81	1.05	1.81	1.05	2.57	1.81	2.57	1.81
2	000011	$1/m$	(0000)	(0001)	(0010)	(0011)	(0100)	(0101)	(0110)	(0111)
			4.09	2.57	2.57	4.09	1.81	0.29	0.29	1.81
			(1000)	(1001)	(1010)	(1011)	(1100)	(1101)	(1110)	(1111)
			2.19	0.67	0.67	2.19	2.95	1.43	1.43	2.95
5	000100	$1/m$	(0000)	(0001)	(0010)	(0011)	(0100)	(0101)	(0110)	(0111)
			4.09	2.57	2.57	4.09	1.81	0.29	0.29	1.81
			(1000)	(1001)	(1010)	(1011)	(1100)	(1101)	(1110)	(1111)
			2.19	0.67	0.67	2.19	2.95	1.43	1.43	2.95
6	000101	$1/m$	(0000)	(0001)	(0010)	(0011)	(0100)	(0101)	(0110)	(0111)
			3.71	2.95	3.71	2.95	1.43	0.67	1.43	0.67
			(1000)	(1001)	(1010)	(1011)	(1100)	(1101)	(1110)	(1111)
			1.81	1.05	1.81	1.05	2.57	1.81	2.57	1.81
7	000110	$1/m$	(0000)	(0001)	(0010)	(0011)	(0100)	(0101)	(0110)	(0111)
			3.52	3.52	3.14	3.14	1.24	1.24	0.86	0.86
			(1000)	(1001)	(1010)	(1011)	(1100)	(1101)	(1110)	(1111)
			1.62	1.62	1.24	1.24	2.38	2.38	2	2
8	000111	$1/m$	(0000)	(0001)	(0010)	(0011)	(0100)	(0101)	(0110)	(0111)
			4.66	2.38	2.76	3.52	2.38	0.19	0.48	1.24
			(1000)	(1001)	(1010)	(1011)	(1100)	(1101)	(1110)	(1111)
			2.76	0.48	0.86	1.62	3.52	1.24	1.62	2.38
9	001000	$1/m$	(0000)	(0001)	(0010)	(0011)	(0100)	(0101)	(0110)	(0111)
			3.52	1.24	1.62	2.38	3.52	1.24	1.62	2.38
			(1000)	(1001)	(1010)	(1011)	(1100)	(1101)	(1110)	(1111)
			3.14	0.86	1.24	2	3.14	0.86	1.24	2
10	001001	$1/m$	(0000)	(0001)	(0010)	(0011)	(0100)	(0101)	(0110)	(0111)
			2.38	2.38	2	2	2.38	2.38	2	2
			(1000)	(1001)	(1010)	(1011)	(1100)	(1101)	(1110)	(1111)
			2	2	1.62	1.62	2	2	1.62	1.62

Table 5. Continued

case	coefficients	w	bounds under $(abcd)$							
			0000)	(0001)	(0010)	(0011)	(0100)	(0101)	(0110)	(0111)
11	001010	$1/m$	2.57	1.81	2.57	1.81	2.57	1.81	2.57	1.81
			(1000)	(1001)	(1010)	(1011)	(1100)	(1101)	(1110)	(1111)
			2.19	1.43	2.19	1.43	2.19	1.43	2.19	1.43
			(0000)	(0001)	(0010)	(0011)	(0100)	(0101)	(0110)	(0111)
12	001011	$1/m$	2.95	1.43	1.43	2.95	2.95	1.43	1.43	2.95
			(1000)	(1001)	(1010)	(1011)	(1100)	(1101)	(1110)	(1111)
			2.57	1.05	1.05	2.57	2.57	1.05	1.05	2.57
			(0000)	(0001)	(0010)	(0011)	(0100)	(0101)	(0110)	(0111)
13	001100	$1/m$	2.95	1.43	1.43	2.95	2.95	1.43	1.43	2.95
			(1000)	(1001)	(1010)	(1011)	(1100)	(1101)	(1110)	(1111)
			2.57	1.05	1.05	2.57	2.57	1.05	1.05	2.57
			(0000)	(0001)	(0010)	(0011)	(0100)	(0101)	(0110)	(0111)
14	001101	$1/m$	2.57	1.81	2.57	1.81	2.57	1.81	2.57	1.81
			(1000)	(1001)	(1010)	(1011)	(1100)	(1101)	(1110)	(1111)
			2.19	1.43	2.19	1.43	2.19	1.43	2.19	1.43
			(0000)	(0001)	(0010)	(0011)	(0100)	(0101)	(0110)	(0111)
15	001110	$1/m$	2.38	2.38	2	2	2.38	2.38	2	2
			(1000)	(1001)	(1010)	(1011)	(1100)	(1101)	(1110)	(1111)
			2	2	1.62	1.62	2	2	1.62	1.62
			(0000)	(0001)	(0010)	(0011)	(0100)	(0101)	(0110)	(0111)
16	001111	$1/m$	3.52	1.24	1.62	2.38	3.52	1.24	1.62	2.38
			(1000)	(1001)	(1010)	(1011)	(1100)	(1101)	(1110)	(1111)
			3.14	0.86	1.24	2	3.14	0.86	1.24	2
			(0000)	(0001)	(0010)	(0011)	(0100)	(0101)	(0110)	(0111)
17	010000	$1/m$	3.71	1.43	1.81	2.57	2.95	0.67	1.05	1.81
			(1000)	(1001)	(1010)	(1011)	(1100)	(1101)	(1110)	(1111)
			3.71	1.43	1.81	2.57	2.95	0.67	1.05	1.81
			(0000)	(0001)	(0010)	(0011)	(0100)	(0101)	(0110)	(0111)
18	010001	$1/m$	2.57	2.57	2.19	2.19	1.81	1.81	1.43	1.43
			(1000)	(1001)	(1010)	(1011)	(1100)	(1101)	(1110)	(1111)
			2.57	2.57	2.19	2.19	1.81	1.81	1.43	1.43
			(0000)	(0001)	(0010)	(0011)	(0100)	(0101)	(0110)	(0111)
19	010010	$1/m$	2.76	2	2.76	2	2	1.24	2	1.24
			(1000)	(1001)	(1010)	(1011)	(1100)	(1101)	(1110)	(1111)
			2.76	2	2.76	2	2	1.24	2	1.24
			(0000)	(0001)	(0010)	(0011)	(0100)	(0101)	(0110)	(0111)
20	010011	$1/m$	3.14	1.62	1.62	3.14	2.38	0.86	0.86	2.38
			(1000)	(1001)	(1010)	(1011)	(1100)	(1101)	(1110)	(1111)
			3.14	1.62	1.62	3.14	2.38	0.86	0.86	2.38
			(0000)	(0001)	(0010)	(0011)	(0100)	(0101)	(0110)	(0111)
21	010100	$1/m$	3.14	1.62	1.62	3.14	2.38	0.86	0.86	2.38
			(1000)	(1001)	(1010)	(1011)	(1100)	(1101)	(1110)	(1111)
			3.14	1.62	1.62	3.14	2.38	0.86	0.86	2.38

case	coefficients	w	bounds under ($abcd$)							
			(0000)	(0001)	(0010)	(0011)	(0100)	(0101)	(0110)	(0111)
			2.76	2	2.76	2	2	1.24	2	1.24
22	010101	$1/m$	(1000)	(1001)	(1010)	(1011)	(1100)	(1101)	(1110)	(1111)
			2.76	2	2.76	2	2	1.24	2	1.24
			(0000)	(0001)	(0010)	(0011)	(0100)	(0101)	(0110)	(0111)
			2.57	2.57	2.19	2.19	1.81	1.81	1.43	1.43
23	010110	$1/m$	(1000)	(1001)	(1010)	(1011)	(1100)	(1101)	(1110)	(1111)
			2.57	2.57	2.19	2.19	1.81	1.81	1.43	1.43
			(0000)	(0001)	(0010)	(0011)	(0100)	(0101)	(0110)	(0111)
			3.71	1.43	1.81	2.57	2.95	0.67	1.05	1.81
24	010111	$1/m$	(1000)	(1001)	(1010)	(1011)	(1100)	(1101)	(1110)	(1111)
			3.71	1.43	1.81	2.57	2.95	0.67	1.05	1.81
			(0000)	(0001)	(0010)	(0011)	(0100)	(0101)	(0110)	(0111)
			4.09	1.81	2.19	2.95	2.57	0.29	0.67	1.43
25	011000	$1/m$	(1000)	(1001)	(1010)	(1011)	(1100)	(1101)	(1110)	(1111)
			2.57	0.29	0.67	1.43	4.09	1.81	2.19	2.95
			(0000)	(0001)	(0010)	(0011)	(0100)	(0101)	(0110)	(0111)
			2.95	2.95	2.57	2.57	1.43	1.43	1.05	1.05
26	011001	$1/m$	(1000)	(1001)	(1010)	(1011)	(1100)	(1101)	(1110)	(1111)
			1.43	1.43	1.05	1.05	2.95	2.95	2.57	2.57
			(0000)	(0001)	(0010)	(0011)	(0100)	(0101)	(0110)	(0111)
			3.14	2.38	3.14	2.38	1.62	0.86	1.62	0.86
27	011010	$1/m$	(1000)	(1001)	(1010)	(1011)	(1100)	(1101)	(1110)	(1111)
			1.62	0.86	1.62	0.86	3.14	2.38	3.14	2.38
			(0000)	(0001)	(0010)	(0011)	(0100)	(0101)	(0110)	(0111)
			3.52	2	2	3.52	2	0.48	0.48	2
28	011011	$1/m$	(1000)	(1001)	(1010)	(1011)	(1100)	(1101)	(1110)	(1111)
			2	0.48	0.48	2	3.52	2	2	3.52
			(0000)	(0001)	(0010)	(0011)	(0100)	(0101)	(0110)	(0111)
			3.52	2	2	3.52	2	0.48	0.48	2
29	011100	$1/m$	(1000)	(1001)	(1010)	(1011)	(1100)	(1101)	(1110)	(1111)
			2	0.48	0.48	2	3.52	2	2	3.52
			(0000)	(0001)	(0010)	(0011)	(0100)	(0101)	(0110)	(0111)
			3.14	2.38	3.14	2.38	1.62	0.86	1.62	0.86
30	011101	$1/m$	(1000)	(1001)	(1010)	(1011)	(1100)	(1101)	(1110)	(1111)
			1.62	0.86	1.62	0.86	3.14	2.38	3.14	2.38
			(0000)	(0001)	(0010)	(0011)	(0100)	(0101)	(0110)	(0111)
			2.95	2.95	2.57	2.57	1.43	1.43	1.05	1.05
31	011110	$1/m$	(1000)	(1001)	(1010)	(1011)	(1100)	(1101)	(1110)	(1111)
			1.43	1.43	1.05	1.05	2.95	2.95	2.57	2.57
			(0000)	(0001)	(0010)	(0011)	(0100)	(0101)	(0110)	(0111)
			4.09	1.81	2.19	2.95	2.57	0.29	0.67	1.43
32	011111	$1/m$	(1000)	(1001)	(1010)	(1011)	(1100)	(1101)	(1110)	(1111)
			2.57	0.29	0.67	1.43	4.09	1.81	2.19	2.95

Table 5. Continued

case	coefficients	w	bounds under $(abcd)$							
			(0000)	(0001)	(0010)	(0011)	(0100)	(0101)	(0110)	(0111)
33	100000	$1/m$	4.09	1.81	2.19	2.95	2.57	0.29	0.67	1.43
			(1000)	(1001)	(1010)	(1011)	(1100)	(1101)	(1110)	(1111)
			2.57	0.29	0.67	1.43	4.09	1.81	2.19	2.95
			(0000)	(0001)	(0010)	(0011)	(0100)	(0101)	(0110)	(0111)
34	100001	$1/m$	2.95	2.95	2.57	2.57	1.43	1.43	1.05	1.05
			(1000)	(1001)	(1010)	(1011)	(1100)	(1101)	(1110)	(1111)
			1.43	1.43	1.05	1.05	2.95	2.95	2.57	2.57
			(0000)	(0001)	(0010)	(0011)	(0100)	(0101)	(0110)	(0111)
35	100010	$1/m$	3.14	2.38	3.14	2.38	1.62	0.86	1.62	0.86
			(1000)	(1001)	(1010)	(1011)	(1100)	(1101)	(1110)	(1111)
			1.62	0.86	1.62	0.86	3.14	2.38	3.14	2.38
			(0000)	(0001)	(0010)	(0011)	(0100)	(0101)	(0110)	(0111)
36	100011	$1/m$	3.52	2	2	3.52	2	0.48	0.48	2
			(1000)	(1001)	(1010)	(1011)	(1100)	(1101)	(1110)	(1111)
			2	0.48	0.48	2	3.52	2	2	3.52
			(0000)	(0001)	(0010)	(0011)	(0100)	(0101)	(0110)	(0111)
37	100100	$1/m$	3.52	2	2	3.52	2	0.48	0.48	2
			(1000)	(1001)	(1010)	(1011)	(1100)	(1101)	(1110)	(1111)
			2	0.48	0.48	2	3.52	2	2	3.52
			(0000)	(0001)	(0010)	(0011)	(0100)	(0101)	(0110)	(0111)
38	100101	$1/m$	3.14	2.38	3.14	2.38	1.62	0.86	1.62	0.86
			(1000)	(1001)	(1010)	(1011)	(1100)	(1101)	(1110)	(1111)
			1.62	0.86	1.62	0.86	3.14	2.38	3.14	2.38
			(0000)	(0001)	(0010)	(0011)	(0100)	(0101)	(0110)	(0111)
39	100110	$1/m$	2.95	2.95	2.57	2.57	1.43	1.43	1.05	1.05
			(1000)	(1001)	(1010)	(1011)	(1100)	(1101)	(1110)	(1111)
			1.43	1.43	1.05	1.05	2.95	2.95	2.57	2.57
			(0000)	(0001)	(0010)	(0011)	(0100)	(0101)	(0110)	(0111)
40	100111	$1/m$	4.09	1.81	2.19	2.95	2.57	0.29	0.67	1.43
			(1000)	(1001)	(1010)	(1011)	(1100)	(1101)	(1110)	(1111)
			2.57	0.29	0.67	1.43	4.09	1.81	2.19	2.95
			(0000)	(0001)	(0010)	(0011)	(0100)	(0101)	(0110)	(0111)
41	101000	$1/m$	3.71	1.43	1.81	2.57	2.95	0.67	1.05	1.81
			(1000)	(1001)	(1010)	(1011)	(1100)	(1101)	(1110)	(1111)
			3.71	1.43	1.81	2.57	2.95	0.67	1.05	1.81
			(0000)	(0001)	(0010)	(0011)	(0100)	(0101)	(0110)	(0111)
42	101001	$1/m$	2.57	2.57	2.19	2.19	1.81	1.81	1.43	1.43
			(1000)	(1001)	(1010)	(1011)	(1100)	(1101)	(1110)	(1111)
			2.57	2.57	2.19	2.19	1.81	1.81	1.43	1.43
			0000)	(0001)	(0010)	(0011)	(0100)	(0101)	(0110)	(0111)
43	101010	$1/m$	2.76	2	2.76	2	2	1.24	2	1.24
			(1000)	(1001)	(1010)	(1011)	(1100)	(1101)	(1110)	(1111)
			2.76	2	2.76	2	2	1.24	2	1.24

case	coefficients	w	bounds under ($abcd$)							
			(0000)	(0001)	(0010)	(0011)	(0100)	(0101)	(0110)	(0111)
			3.14	1.62	1.62	3.14	2.38	0.86	0.86	2.38
44	101011	1/m	(1000)	(1001)	(1010)	(1011)	(1100)	(1101)	(1110)	(1111)
			3.14	1.62	1.62	3.14	2.38	0.86	0.86	2.38
			(0000)	(0001)	(0010)	(0011)	(0100)	(0101)	(0110)	(0111)
			3.14	1.62	1.62	3.14	2.38	0.86	0.86	2.38
45	101100	1/m	(1000)	(1001)	(1010)	(1011)	(1100)	(1101)	(1110)	(1111)
			3.14	1.62	1.62	3.14	2.38	0.86	0.86	2.38
			(0000)	(0001)	(0010)	(0011)	(0100)	(0101)	(0110)	(0111)
			2.76	2	2.76	2	2	1.24	2	1.24
46	101101	1/m	(1000)	(1001)	(1010)	(1011)	(1100)	(1101)	(1110)	(1111)
			2.76	2	2.76	2	2	1.24	2	1.24
			(0000)	(0001)	(0010)	(0011)	(0100)	(0101)	(0110)	(0111)
			2.57	2.57	2.19	2.19	1.81	1.81	1.43	1.43
47	101110	1/m	(1000)	(1001)	(1010)	(1011)	(1100)	(1101)	(1110)	(1111)
			2.57	2.57	2.19	2.19	1.81	1.81	1.43	1.43
			(0000)	(0001)	(0010)	(0011)	(0100)	(0101)	(0110)	(0111)
			3.71	1.43	1.81	2.57	2.95	0.67	1.05	1.81
48	101111	1/m	(1000)	(1001)	(1010)	(1011)	(1100)	(1101)	(1110)	(1111)
			3.71	1.43	1.81	2.57	2.95	0.67	1.05	1.81
			(0000)	(0001)	(0010)	(0011)	(0100)	(0101)	(0110)	(0111)
			3.52	1.24	1.62	2.38	3.52	1.24	1.62	2.38
49	110000	1/m	(1000)	(1001)	(1010)	(1011)	(1100)	(1101)	(1110)	(1111)
			3.14	0.86	1.24	2	3.14	0.86	1.24	2
			(0000)	(0001)	(0010)	(0011)	(0100)	(0101)	(0110)	(0111)
			2.38	2.38	2	2	2.38	2.38	2	2
50	110001	1/m	(1000)	(1001)	(1010)	(1011)	(1100)	(1101)	(1110)	(1111)
			2	2	1.62	1.62	2	2	1.62	1.62
			(0000)	(0001)	(0010)	(0011)	(0100)	(0101)	(0110)	(0111)
			2.57	1.81	2.57	1.81	2.57	1.81	2.57	1.81
51	110010	1/m	(1000)	(1001)	(1010)	(1011)	(1100)	(1101)	(1110)	(1111)
			2.19	1.43	2.19	1.43	2.19	1.43	2.19	1.43
			(0000)	(0001)	(0010)	(0011)	(0100)	(0101)	(0110)	(0111)
			2.95	1.43	1.43	2.95	2.95	1.43	1.43	2.95
52	110011	1/m	(1000)	(1001)	(1010)	(1011)	(1100)	(1101)	(1110)	(1111)
			2.57	1.05	1.05	2.57	2.57	1.05	1.05	2.57
			(0000)	(0001)	(0010)	(0011)	(0100)	(0101)	(0110)	(0111)
			2.95	1.43	1.43	2.95	2.95	1.43	1.43	2.95
53	110100	1/m	(1000)	(1001)	(1010)	(1011)	(1100)	(1101)	(1110)	(1111)
			2.57	1.05	1.05	2.57	2.57	1.05	1.05	2.57
			(0000)	(0001)	(0010)	(0011)	(0100)	(0101)	(0110)	(0111)
			2.57	1.81	2.57	1.81	2.57	1.81	2.57	1.81
54	110101	1/m	(1000)	(1001)	(1010)	(1011)	(1100)	(1101)	(1110)	(1111)
			2.19	1.43	2.19	1.43	2.19	1.43	2.19	1.43

Table 5. Continued

case	coefficients	w	bounds under ($abcd$)							
			(0000)	(0001)	(0010)	(0011)	(0100)	(0101)	(0110)	(0111)
55	110110	$1/m$	2.38	2.38	2	2	2.38	2.38	2	2
			(1000)	(1001)	(1010)	(1011)	(1100)	(1101)	(1110)	(1111)
			2	2	1.62	1.62	2	2	1.62	1.62
			(0000)	(0001)	(0010)	(0011)	(0100)	(0101)	(0110)	(0111)
56	110111	$1/m$	3.52	1.24	1.62	2.38	3.52	1.24	1.62	2.38
			(1000)	(1001)	(1010)	(1011)	(1100)	(1101)	(1110)	(1111)
			3.14	0.86	1.24	2	3.14	0.86	1.24	2
			(0000)	(0001)	(0010)	(0011)	(0100)	(0101)	(0110)	(0111)
57	111000	$1/m$	4.66	2.38	2.76	3.52	2.38	0.19	0.48	1.24
			(1000)	(1001)	(1010)	(1011)	(1100)	(1101)	(1110)	(1111)
			2.76	0.48	0.86	1.62	3.52	1.24	1.62	2.38
			(0000)	(0001)	(0010)	(0011)	(0100)	(0101)	(0110)	(0111)
58	111001	$1/m$	3.52	3.52	3.14	3.14	1.24	1.24	0.86	0.86
			(1000)	(1001)	(1010)	(1011)	(1100)	(1101)	(1110)	(1111)
			1.62	1.62	1.24	1.24	2.38	2.38	2	2
			(0000)	(0001)	(0010)	(0011)	(0100)	(0101)	(0110)	(0111)
59	111010	$1/m$	3.71	2.95	3.71	2.95	1.43	0.67	1.43	0.67
			(1000)	(1001)	(1010)	(1011)	(1100)	(1101)	(1110)	(1111)
			1.81	1.05	1.81	1.05	2.57	1.81	2.57	1.81
			(0000)	(0001)	(0010)	(0011)	(0100)	(0101)	(0110)	(0111)
60	111011	$1/m$	4.09	2.57	2.57	4.09	1.81	0.29	0.29	1.81
			(1000)	(1001)	(1010)	(1011)	(1100)	(1101)	(1110)	(1111)
			2.19	0.67	0.67	2.19	2.95	1.43	1.43	2.95
			(0000)	(0001)	(0010)	(0011)	(0100)	(0101)	(0110)	(0111)
61	111100	$1/m$	4.09	2.57	2.57	4.09	1.81	0.29	0.29	1.81
			(1000)	(1001)	(1010)	(1011)	(1100)	(1101)	(1110)	(1111)
			2.19	0.67	0.67	2.19	2.95	1.43	1.43	2.95
			(0000)	(0001)	(0010)	(0011)	(0100)	(0101)	(0110)	(0111)
62	111101	$1/m$	3.71	2.95	3.71	2.95	1.43	0.67	1.43	0.67
			(1000)	(1001)	(1010)	(1011)	(1100)	(1101)	(1110)	(1111)
			1.81	1.05	1.81	1.05	2.57	1.81	2.57	1.81
			(0000)	(0001)	(0010)	(0011)	(0100)	(0101)	(0110)	(0111)
63	111110	$1/m$	3.52	3.52	3.14	3.14	1.24	1.24	0.86	0.86
			(1000)	(1001)	(1010)	(1011)	(1100)	(1101)	(1110)	(1111)
			1.62	1.62	1.24	1.24	2.38	2.38	2	2
			(0000)	(0001)	(0010)	(0011)	(0100)	(0101)	(0110)	(0111)
64	111111	$1/m$	4.66	2.38	2.76	3.52	2.38	0.19	0.48	1.24
			(1000)	(1001)	(1010)	(1011)	(1100)	(1101)	(1110)	(1111)
			2.76	0.48	0.86	1.62	3.52	1.24	1.62	2.38

REFERENCES

[1] Fan and P. Z. Darnell M., *Sequence design for communications applications,* John Wiley & Sons, Ltd., 1996.

[2] Chen H. H., *The next generation CDMA technologies,* John Wiley & Sons, Ltd., 2007.

[3] Litsyn S., *Peak power control in multicarrier communications,* Cambridge University Press, 2007.

[4] Davis J. A. and Jedwab J., "Peak-to-mean power control in OFDM, Golay complementary sequences, and Reed-Muller codes," *IEEE Transactions on Information Theory,* vol. 45, no.7, pp. 2397-2417, Nov. 1999.

[5] Fiedler F., Jedwab J. and Parker M. G., "A multi-dimensional approach to the construction and enumeration of Golay complementary sequences," *J. Comb. Theory, Series A*, vol. 115, pp. 753-776, 2008.

[6] Fiedler F., Jedwab J. and Parker M. G., "A framework for the construction of Golay sequences," *IEEE Trans. Inf. Theory*, vol. 54, no.7, pp.3114-3129, July 2008.

[7] Anand M. and Vijay Kumar P., "Low-Correlation Sequences Over the QAM Constellation," *IEEE Trans. Inf. Theory,* vol. 54, no. 2, pp.791-810, Feb. 2008.

[8] Chong C. V., Venkataramani R. and Tarokh V., "A new construction of 16-QAM Golay complementary sequences," *IEEE Trans. Inf. Theory,* vol. 49, no.11, pp. 2953-2959, Nov. 2003.

[9] Li Y., "A construction of general QAM Golay complementary sequences," *IEEE Trans. Inf. Theory,* vol. 56, no. 11, pp.5765-5771, Nov. 2010.

[10] Zeng Y, Zhang L. S., Zeng F. X., He X. P., Xuan G. X. and Zhang Z. Y., "New QAM complementary sequences for control of peak envelope power of OFDM signals," *IEEE Access,* vol. 7, pp. 89901-89912, 2019.

[11] Zeng F. X., He X. P., Xuan G. X., Zhang Z. Y., li G. J., Peng Y. N., Lu S. and Yan L., "More general QAM complementary sequences," *IEICE Transactions on Fundamentals of Electronics, Communications and Computer Sciences,* vol. E101-A, no. 12, pp.2409-2414, Dec. 2018.

[12] Zeng F. X., "A sufficient condition producing 16-QAM Golay complementary sequences", *IEEE Commun. Lett.*, vol. 18, no. 11, pp.1875-1878, Nov. 2014.

[13] Liu Z. L., Li Y. and Guan Y. L., "New constructions of general QAM Golay complementary sequences," *IEEE Trans. Inf. Theory,* vol.59, no. 11, pp. 7684-7692, Nov. 2013.

[14] Liu X. and Sethumadhvavn C., Method and apparatus for transmitting high-level QAM optical signals with binary drive signals, Appl. No. 13/340,916, US patent, July 2013.

[15] Zeng F. X., Zeng Y., Zhang L. S., He X. P., Xuan G. X. and Zhang Z. Y., "A sufficient condition for general QAM complementary sequence pairs," *Canadian Journal of Electrical and Computer Engineer,* vol. 43, no. 1, pp.43-56, 2020. To appear.

[16] Zeng F. X., Zeng X. P. and Zhang Z. Y. "Novel 16-QAM complementary sequences," *IEICE Transactions on Fundamentals of Electronics, Communications and Computer Sciences,* vol. E97-A, no. 7, July 2014, pp.1631-1634.

INDEX

NONLINEAR INTEGRAL EQUATIONS ON TIME SCALES

AUTHORS: Svetlin G. Georgiev and Inci M. Erhan

SERIES: Theoretical and Applied Mathematics

BOOK DESCRIPTION: This book describes experiments that have proved that gravity, velocity, and acceleration slow time. This book contains different analytical and numerical methods for investigation of nonlinear integral equations on time scales.

HARDCOVER ISBN: 978-1-53615-021-6
RETAIL PRICE: $230

WAVELETS: PRINCIPLES, ANALYSIS AND APPLICATIONS

EDITOR: Joseph Burgess

SERIES: Theoretical and Applied Mathematics

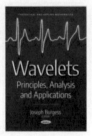

BOOK DESCRIPTION: In this book, the authors report the results obtained by the application of wavelet analysis to two physics experiments: the motion of variable mass pendulum and the motion of variable length pendulum.

SOFTCOVER ISBN: 978-1-53613-374-5
RETAIL PRICE: $95